USING PHYSICAL SCIENCE
GADGETS & GIZMOS
GRADES 6–8
PHENOMENON-BASED LEARNING

National Science Teachers Association

Arlington, Virginia

National Science Teachers Association

Claire Reinburg, Director
Wendy Rubin, Managing Editor
Andrew Cooke, Senior Editor
Amanda O'Brien, Associate Editor
Amy America, Book Acquisitions Coordinator

ART AND DESIGN
Will Thomas Jr., Director
Joe Butera, Senior Graphic Designer, cover and
    interior design

PRINTING AND PRODUCTION
Catherine Lorrain, Director

NATIONAL SCIENCE TEACHERS ASSOCIATION
David L. Evans, Executive Director
David Beacom, Publisher

1840 Wilson Blvd., Arlington, VA 22201
www.nsta.org/store
For customer service inquiries, please call 800-277-5300.

Copyright © 2014 by the National Science Teachers Association.
All rights reserved. Printed in the United States of America.
17 16 15 14   4 3 2 1

Cataloging-in-Publication Data are available from the Library of Congress.
ISBN: 978-1-936959-37-2
e-ISBN: 978-1-938946-61-5

C O N T E N T S

1

WAVE MOTION AND SOUND   1

2

VISIBLE LIGHT AND COLORS   17

ABOUT THE AUTHORS

MATTHEW BOBROWSKY, PHD

Dr. Matt Bobrowsky has been involved in scientific research and science education for several decades. For four years, he served as Director of the Physics Demonstration Facility at the University of Maryland—a collection of over 1,600 science demonstrations. Also at the University of Maryland, Matt was selected as a Faculty Mentor for the Fulbright Distinguished International Teachers Program, where he met Mikko Korhonen.

Matt has taught physics, astronomy, and astrobiology both in the classroom and online. He has written K–12 science curricula and serves on the Science Advisory Committee for the Howard County Public School System in Maryland. Matt has conducted countless professional development workshops for science teachers and special presentations for students, speaking on a variety of topics beyond physics, such as the scale of the universe, life in the universe, misconceptions about science among students and the public, the process of science, and science versus pseudoscience. He is often asked to be an after-dinner speaker or keynote speaker at special events. Matt is a "Nifty Fifty" speaker for the USA Science & Engineering Festival and a Shapley Lecturer for the American Astronomical Society. Matt has received a number of awards for teaching excellence from the University of Maryland, including the Stanley J. Drazek Teaching Excellence Award (given to the top 2 instructors out of ~800) and the Board of Regents' Faculty Award for Excellence in Teaching (given to the top 3 instructors out of ~7,000). Matt's teaching is always innovative because he uses pedagogical techniques that are based on current science education research and known to be effective.

In his research, Matt has been involved in both theoretical and observational astronomy. He developed computer models of planetary nebulae—clouds of gas expanding outward from aging stars—and has observed them with telescopes on the ground as well as with the Hubble Space Telescope. One of the planetary nebulae that Matt investigated is the Stingray Nebula, which he discovered using Hubble.

MIKKO KORHONEN

Mikko Korhonen obtained a master's degree from Tampere Technical University in Finland, where he studied physics, mathematics, and pedagogics. Since then, he has been teaching physics, mathematics, and computer science at various schools in Finland. He has also developed a number of educational programs that brought some of his students to top scientific facilities in the world, including the Nordic Optical Telescope (NOT) observatory in La Palma, Spain, the CERN laboratory at the Franco-Swiss border, and the LATMOS laboratory in France. Most recently, some of his students have attended the Transatlantic Science School, which Mikko founded.

Mikko has written numerous other educational publications, including a book of physics experiments, manuals of physics problems with answers, an article on mathematics and logic for computer science, and two books with Jukka Kohtamäki on using toys to teach physics, one at the middle school and one at the high school level. (This book is an adaptation of the Finnish version of the middle school book.)

Mikko has obtained numerous grants and awards for his school and students, including awards from the NOT science school and the Viksu science competition prize, as well as individual grants from the Finnish National Board of Education and the Technology Industries of Finland Centennial Foundation, and grants for his "physics toys" project. His students are also award winners in the Finnish National Science Competition. Mikko received one of the Distinguished Fulbright Awards in Teaching, which brought him to the University of Maryland for a semester, where he worked with Matt Bobrowsky. Most recently, Mikko received the award of Distinguished Science Teacher in 2013 by the Technology Industries of Finland Centennial Foundation.

JUKKA KOHTAMÄKI

Jukka Kohtamäki obtained his master of science degree from Tampere University of Technology in Finland and since then has been teaching grades 5–9 at the Rantakylä Comprehensive School, one of the largest comprehensive schools in Finland. Jukka has participated in long-term professional development teaching projects and projects involving the use of technology in learning, as well as workshops that he and Mikko Korhonen conducted for Finnish science teachers. His writing includes teaching materials for physics and computer science, and he has written two books with Mikko on using toys to teach physics, one at the middle school and one at the high school level. (This book is an adaptation of the Finnish version of the middle school book.)

Jukka is a member of the group under the National Board of Education that is writing the next physics curriculum in Finland. He is also participating in writing curricula in chemistry and natural science (which is a combination of biology, geology, physics, chemistry, and health education). His goals are to get students engaged in lessons, to have them work hands on and minds-on, to encourage creativity in finding solutions, and to get students to discuss natural phenomena using the "language of physics." In 2013, Jukka received the Distinguished Science Teacher Award from the Technology Industries of Finland Centennial Foundation.

"The most beautiful thing
we can experience is the
mysterious. It is the source
of all true art and science."

— Albert Einstein

AN INTRODUCTION TO PHENOMENON-BASED LEARNING

TO THE STUDENT

In 1931 Albert Einstein wrote, "The most beautiful thing we can experience is the mysterious. It is the source of all true art and science." Keep this in mind as we introduce you to phenomenon-based learning, a learning approach in which you start by observing a natural phenomenon—in some cases just a simple toy—and then build scientific models and theories based on your observations.

The goal is for you to first watch something happen and then become curious enough to find out why. You will experiment with some simple gizmos and think about them from different perspectives. Developing a complete understanding of a concept might take a number of steps, with each step providing a deeper understanding of the topic. In some cases, you will need to do further research on your own to understand certain terms and concepts. Like real scientists, you can also get help from (and provide help to) collaborators. This book's approach to learning is based on curiosity and creativity—a fun way to learn!

TO THE TEACHER

The pedagogical approach in this book is called phenomenon-based learning (PBL), meaning learning is built on observations of real-world phenomena—in this case of some fun toys or gadgets. The method also uses peer instruction, which research has shown results in more learning than traditional lectures (Crouch and Mazur 2001). In the PBL approach, students work and explore in groups: Exercises are done in groups, and students' conclusions are also drawn in groups. The teacher guides and encourages the groups and, at the end, verifies the conclusions. With the PBL strategy, the concepts and the phenomena are approached from different angles, each adding a piece to the puzzle with the goal of developing a picture correctly portraying the real situation.

The activities in this book can be used for various purposes. The introductions and the questions can be used as the basis for discussions with the groups before the students use the gizmos, that is, as a motivational tool. For example, you can ask where we see or observe the phenomenon in everyday life, what the students know about the matter prior to conducting the activities, and so on.

PBL is not so much a teaching method as it is a route to grasping the big picture. It contains some elements that you may have seen in inquiry-based, problem-based, or project-based learning, combined with hands-on activities. In traditional physics teaching, it's common to divide phenomena into small, separate parts and discuss them as though there is no connection among them (McNeil 2013). In our PBL approach, we don't artificially create boundaries within phenomena. Rather, we try to look at physical phenomena very broadly.

PBL encourages students to not just think about what they have learned but to also reflect on how they acquired that knowledge. What mental processes did they go through while exploring a phenomenon and figuring out what was happening? PBL very much lends itself to a K-W-L approach (what we **K**now, what we **W**ant to know, and what we have **L**earned). K-W-L can be enhanced by adding an H for "**H**ow we learned it" because once we understand that, we can apply those same learning techniques to other situations.

When you first look at this book, it might seem as if there is not very much textual material. That was intentional. The idea is to have more thinking by the students and less lecturing by the teacher. It is also important to note that the process of thinking and learning is not a race. To learn and really get the idea, students need to take time to think … and then think some more—so be sure to allow sufficient time for the cognitive processes to occur. For example, the very first experiment (using a tuning fork) can be viewed in two seconds, but in order for students to think about the phenomenon and really get the idea, they need to discuss the science with other group members, practice using the "language of science," and internalize the science involved—which might take 20 minutes. During this time, the students may also think of real-life situations in which the phenomenon plays a significant role, and these examples can be brought up later during discussions as an entire class.

"Most of the time my students didn't need me; they were just excited about a connection or discovery they made and wanted to show me."

—*Jamie Cohen (2014)*

LEARNING GOALS AND ASSESSMENT

The most important learning goal is for students to learn to think about problems and try a variety of approaches to solve them. Nowadays, most students just wait for the teacher to state the answer. The aim here is for students to enjoy figuring out what's going on and to be creative and innovative. Combining this with other objectives, a list of learning goals might look something like this:

By the end of these lessons, students will

- think about problems from various angles and try different strategies;

- demonstrate process skills, working logically and consistently;

- collaborate with others to solve problems;

- use the language of physical science (and science in general);

- reflect on the thinking processes that helped them to acquire new knowledge and skills in physical science; and

- view physical science as interesting and fun.

You will also notice that there are no formal quizzes or rubrics included. There are other ways to evaluate students during activities such as these. First, note that the emphasis is not on getting the "right" answer. Teachers should not simply provide the answer or an easy way out—that would not allow students to learn how science really works. When looking at student answers, consider the following: Are the students basing their conclusions on evidence? Are they sharing their ideas with others in their group? Even if a student has the wrong idea, if she or he has evidential reasons for that idea, then that student has the right approach. After all members of a group are in agreement and tell you, the teacher, what they think is happening, you can express doubt or question the group's explanation, making the students describe their evidence and perhaps having them discuss it further among themselves. Student participation as scientific investigators and their ability to give reasons for their explanations will be the key indicators that the students understand the process of science.

The PBL approach lends itself well to having students keep journals of their activities. Students should write about how they are conducting their experiment (which might differ from one group to another), ideas they have related to the phenomenon under investigation (including both correct and incorrect ideas), what experiments or observations showed the incorrect ideas to be wrong, answers to the questions supplied for each exploration, and what they learned as a result of the activity. The teacher can encourage students to form a mental model—perhaps expressed as a drawing—of how the phenomenon works and why. Then they can update this model in the course of their investigations. Students might also want to make a video of the experiment. This can be used for later reference as well as to show family and friends. Wouldn't it be great if we can get students talking about science outside the classroom?

A few of the questions asked of the students will be difficult to answer. Here again, students get a feel for what it's like to be a real scientist exploring uncharted territory. A student might suggest an incorrect explanation. Other students in the group might offer a correction, or if no one does, perhaps

further experimentation, along with guidance from the teacher, will lead the students on the right course. Like scientists, the students can do a literature search (usually a web search now) to see what others know about the phenomenon. Thus there are many ways for a misconception to get dispelled in a way that will result in more long-term understanding than if the students were simply told the answer. Guidance from the teacher could include providing some ideas about what to observe when doing the experiment or giving some examples from other situations in which the same phenomenon takes place. Although many incorrect ideas will not last long in group discussions, the teacher should actively monitor the discussions, ensuring that students do not get too far off track and are on their way to achieving increased understanding. We've provided an analysis of the science behind each exploration to focus your instruction.

By exploring first and getting to a theoretical understanding later, students are working like real scientists. When scientists investigate a new phenomenon, they aren't presented with an explanation first—they have to figure it out. And that's what the students do in PBL. Real scientists extensively collaborate with one another; and that's exactly what the students do here as well—work in groups. Not all terms and concepts are extensively explained; that's not the purpose of this book. Again, like real scientists the students can look up information as needed in, for example, a traditional physics textbook. What we present here is the PBL approach, in which students explore first and are inspired to pursue creative approaches to answers—and have fun in the process!

PBL IN FINLAND

The Finnish educational system came into the spotlight after the Programme of International

AN INTRODUCTION TO
PHENOMENON-BASED LEARNING

Student Assessment (PISA) showed that Finnish students were among the top in science literacy proficiency levels. In 2009, Finland ranked second in science and third in reading out of 74 countries. (The United States ranked 23rd and 17th, respectively.) In 2012, Finland ranked 5th in science. (The U.S. was 28th.) Finland remains #1 in science among member nations of the Organization for Economic Co-operation and Development (OECD). Finland is now seen as a major international leader in education, and its performance has been especially notable for its significant consistency across schools. No other country has so little variation in outcomes among schools, and the gap within schools between the top- and bottom-achieving students is quite small as well. Finnish schools seem to serve all students well, regardless of family background or socioeconomic status. Recently, U.S. educators and political leaders have been traveling to Finland to learn the secret of their success.

The PBL approach is one that includes progressive inquiry, problem-based learning, project-based learning, and in Finland at least, other methods at the teachers' discretion. The idea is to teach bigger concepts and useful thinking skills rather than asking students to memorize everything in a textbook.

AUTHORS' USE OF GADGETS AND GIZMOS

One of the authors (M.B.) has been using gizmos as the basis of teaching for many years. He also uses them for illustrative purposes in public presentations and school programs. The other two authors (M.K. and J.K.) have been using PBL—and the materials in this book—to teach in Finland. Their approach is to present physics phenomena to students so that they can build ideas and an understanding of the topic by themselves, in small groups. Students progress from thinking to understanding to explaining. For each phenomenon there are several different viewpoints from which the student can develop a big-picture understanding as a result of step-by-step exploration. The teacher serves only as a guide who leads the student in the right direction. PBL is an approach that is not only effective for learning but is also much more fun and interesting for both the teacher and the students.

SAFETY NOTES

Doing science through hands-on, process, and inquiry-based activities or experiments helps to foster the learning and understanding of science. However, in order to make for a safer experience, certain safety procedures must be followed. Throughout this book, there are a series of safety notes that help make PBL a safer learning experience for students and teachers. In most cases, eye protection is required. Safety glasses or safety goggles noted must meet the ANSI Z87.1 safety standard. For additional safety information, check out NSTA's "Safety in the Science Classroom" at *www.nsta.org/pdfs/SafetyInTheScienceClassroom.pdf*. Additional information on safety can be found at the NSTA Safety Portal at *www.nsta.org/portals/safety.aspx*.

REFERENCES

Cohen, J. 2014. 18 Ways to engage your students by teaching less and learning more with rap genius. *http://poetry.rapgenius.com/Mr-cohen-18-ways-to-engage-your-students-by-teaching-less-and-learning-more-with-rap-genius-lyrics*

Crouch, C. H., and E. Mazur. 2001. Peer instruction: Ten years of experience and results. *American Journal of Physics* 69 (9): 970–977.

McNeil, L. E. 2013. Transforming introductory physics teaching at UNC-CH. University of North Carolina at Chapel Hill. *http://user.physics.unc.edu/~mcneil/physicsmanifesto.html*.

Moursund, D. 2013. Problem-based learning and project-based learning. University of Oregon. *http://pages.uoregon.edu/moursund/Math/pbl.htm*

ADDITIONAL RESOURCES

Bobrowsky, M. 2007. *The process of science: and its interaction with non-scientific ideas*. Washington, DC: American Astronomical Society. *http://aas.org/education/The_Process_of_Science*.

Champagne, A. B., R. F. Gunstone, and L. E. Klopfer. 1985. Effecting changes in cognitive structures among physics students. In *Cognitive structure and conceptual change*, ed. H. T. West and A. L. Pines, 163–187. Orlando, FL: Academic Press.

Chi, M. T. H., and R. D. Roscoe. 2002. The processes and challenges of conceptual change. In *Reconsidering conceptual change: Issues in theory and practice*, ed. M. Limón and L. Mason, 3–27. Boston: Kluwer Academic Publishers.

Clement, J. 1982. Students' preconceptions in introductory mechanics. *American Journal of Physics* 50 (1): 66–71.

Clement, J. 1993. Using bridging analogies and anchoring intuitions to deal with students' preconceptions in physics. *Journal of Research in Science Teaching* 30 (10): 1241–1257.

Enger, S. K., and R. E. Yager. 2001. *Assessing student understanding in science: A standards-based K–12 handbook*. Thousand Oaks, CA: Corwin Press.

Gray, K. E., W. K. Adams, C. E. Wieman, and K. K. Perkins. 2008. Students know what physicists believe, but they don't agree: A study using the CLASS survey. *Physics Review Special Topic– Physics Education Research* 4: 020106

Jacobs, H. H., ed. 2010. *Curriculum 21: Essential education for a changing world*. Alexandria, VA: ASCD.

Jones, L. 2007. *The student-centered classroom*. New York, Cambridge University Press. *www.cambridge.org/other_files/downloads/esl/booklets/Jones-Student-Centered.pdf.*

Lucas, A. F. 1990. Using psychological models to understand student motivation. In *The changing face of college teaching: New directions for teaching and learning*, no. 42, ed. M. D. Svinicki. San Francisco: Jossey-Bass

McDade, M. 2013. Children learn better when they figure things out for themselves: Brandywine professor's research published in journal. PennState News. *http://news.psu.edu/story/265620/2013/02/21/society-and-culture/children-learn-better-when-they-figure-things-out*

McTighe, J., and G. Wiggins. 2013. *Essential questions: Opening doors to student understanding*. Alexandria, VA: ASCD.

Meadows, D. H. 2008. *Thinking in systems: A primer*. White River Junction, VT: Chelsea Green Publishing.

National Research Council. 2000. *How people learn: Brain, mind, experience, and school*. Washington, DC: National Academies Press.

National Research Council (NRC). 2012. *A framework for K–12 science education: Practices, crosscutting concepts, and core ideas*. Washington, DC: National Academies Press.

Nissani, M. 1997. Can the persistence of misconceptions be generalized and explained? *Journal of Thought* 32: 69–76. *www.is.wayne.edu/mnissani/pagepub/theory.htm*.

Verley, J. D. 2008. Physics graduate students' perceptions of the value of teaching, PhD diss., University of Wyoming. *http://udini.proquest.com/view/physics-graduate-students-goid:304450532*.

WAVE MOTION AND SOUND

Sound is one example of wave motion. It can arise from the membrane of a drum, guitar strings, or vocal folds vibrating. (Vocal folds are sometimes referred to as "vocal cords," but that is not a good description because they are not cords—like vibrating strings—but more like folds of skin.) Our ability to hear sound is based on the principle of resonance. The eardrum starts to vibrate with the same frequency as the sound waves. Then, the bones in the middle ear transmit the vibration to nerves, and nerves take the signals to the brain. In the following activities, you will learn about standing waves, resonance, and an interesting wave phenomenon known as the Doppler effect.

Sound needs a medium to move through. A medium is any "stuff" that has parts that can vibrate as the result of a sound wave. Air or water can serve as a medium for sound waves. In gases and liquids, the sound waves are *longitudinal waves* (also called *compression waves*). Another type of wave motion is *transverse wave motion*. Sound in solids can propagate as transverse waves as well as longitudinal waves. The speed of sound depends on the medium. With these activities, you will explore what sound actually is and how it is created and heard.

KEYWORDS
Knowing these terms will help you to enjoy the explorations.

wave motion

sound wave

frequency

wavelength

resonance

amplitude

medium

echo

Exploration

FIGURE 1.1: Tuning fork

SAFETY NOTE

Make sure there is nothing breakable in the vicinity when you strike the tuning fork.

GET IN TUNE

Tuning forks (Figure 1.1) can be used to tune a musical instrument and make sure that exactly the right note comes out. A tuning fork produces a single note with one specific pitch, which allows one instrument to produce a sound at the same pitch as the corresponding note played on another instrument. When this is done, all the instruments in an orchestra play in harmony.

In addition to the fork itself, you will also need a small amount of finely ground sugar, a glass of water, and something made of metal such as the leg of a chair.

Procedure

1. Hold the tuning fork by its handle.

2. Before each of the next steps, slap the tuning fork against something solid, such as metal or wood, so it makes a sound.

3. Slowly put the prongs of the fork into a pile of the sugar. Explore what happens and explain your observations.

4. Gently touch something made of metal with the prongs of the fork.

5. Slowly dip the prongs of the tuning fork into water.

Questions

• What did you observe during the experiments?

• How does a tuning fork create a sound?

• How does the sound travel through the air to your ear?

• When the tuning fork is creating a sound, touch your forehead with its handle. What is the sound doing now?

• Cover your ear with your hand. Then, touch your knuckles with the handle of the tuning fork. What do you notice about the sound? Explain.

STANDING WAVE

With the Standing Wave Apparatus (Figure 1.2) you will learn about the formation of standing waves, and you can discuss the nodes and antinodes formed in the thread.

Procedure

1. Insert the AA battery into the motor.

2. Hold the thread in place at the end opposite the motor.

3. Write in your notebook what you observe about the standing wave and its shape. You might want to draw a picture as well.

Questions

•   When the apparatus is working, a standing wave is formed. Make notes on your notebook about the standing wave and its shape.

•   How many nodes and antinodes are formed?

•   How can you change the number of nodes and antinodes?

•   In what instruments are similar standing waves created?

▼
SAFETY NOTE

Wear safety glasses or goggles.

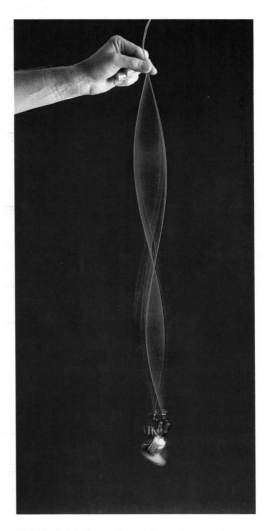

FIGURE 1.2: Standing Wave Apparatus

FIGURE 1.3: Sound Pipe

SOUND PIPE

The Sound Pipe (Figure 1.3) can be used to demonstrate the formation of sound, standing wave motion, and resonance. It can also be used when studying the difference in pressure between the two ends of the pipe. (For more about pressure, see Chapter 4.)

Procedure

Without hitting anything with the pipe, try to get the pipe to make some sound. *Hint:* Make air flow through the pipe.

Questions

- Explain how you created the sound.

- Did you try twirling the pipe around above your head? If not, find a clear place and do it now.

- How can you create a higher pitch with the pipe?

- What happens to the sound if you block one end of the pipe? Why?

- In the introduction of this chapter, it was mentioned that sound is all about vibrations and waves. What is vibrating in the tube?

▼
SAFETY NOTES
- Wear safety glasses or goggles.
- Only swing the pipe in an open area clear of lab equipment and occupants.

MUSIC BOX

The Music Box (Figure 1.4) can be used to study sound resonance. Resonance is an essential phenomenon in musical instruments. In addition, resonance can be used when studying the transfer of energy with sound.

Procedure

1. Figure out how the gadget works.

2. After winding up the music box, place the music box on different surfaces, such as a blackboard, table, or window.

3. While the music box is playing on a table, put your ear against the table.

4. Try building a simple ear phone: Put the music box on your elbow while your index finger is in your ear.

Questions

-   Briefly describe how the gadget works.

-   How is the sound different when the music box is placed on different surfaces?

-   Why is the sound sometimes louder and sometimes less loud?

-   What kinds of surfaces and objects make the sound loudest?

-   Resonance is a phenomenon in which a certain frequency reinforces the vibration of the system. Explain how resonance plays a role here.

-   Describe your observations when the box is played on the table and you put your ear against the table. Explain.

FIGURE 1.4: Music Box

▼
SAFETY NOTE

Make sure there is nothing breakable in the vicinity.

▼
SAFETY NOTES

- Wear safety glasses or goggles.

- Only throw the ball in an open area clear of lab equipment and occupants.

DOPPLER BALL

The Doppler Ball (Figure 1.5) can be used to study the Doppler effect. This effect is the reason that a car sounds different when it's coming toward you compared with when it is moving away from you.

FIGURE 1.5: Doppler Ball

Procedure

1. Assemble the Doppler Ball: Insert the battery, and if it's not already there, place the sound source in the ball.

2. Throw the ball to a friend on the other side of the classroom. Then, have them toss it to someone else. Listen how the sound is different when the ball is in motion.

3. Next, put the ball in a strong bag and swing the ball around. Listen. Take notes on your observations.

Questions

- What is the gadget, and how does it work?

- What did you notice while throwing the ball around?

- You already know that sound is actually compression waves. The higher the frequency of the sound waves, the higher the pitch of the sound. With these facts in mind, create a possible explanation of the phenomenon observed here. Why did you hear different frequencies while the ball was flying? *Hint*: Drawing a sketch of the sound wave might help you to come up with an idea.

- How does the ball sound when it is swung around in a bag?

- Now, how do you explain the changes in frequency in terms of sound waves?

Analysis

GET IN TUNE

Sound is all about vibrations. Vibrating objects such as the tuning fork (Figure 1.6) create sound. When you hit the tuning fork against something solid, it starts to vibrate, which makes the molecules of air move and creates compression waves in the air. These waves cause your eardrums to vibrate, resulting in you hearing a sound. When the tuning fork is almost touching your bones (like when it was pressed against your forehead), the sound you hear is louder because the vibration travels better in solid media and therefore makes the eardrums vibrate more.

FIGURE 1.6: Tuning fork

STANDING WAVE

The Standing Wave Apparatus (Figure 1.7) allows you to explore an important phenomenon in music. In musical instruments, certain parts such as the strings of a guitar or the air in a wind instrument start to vibrate at their characteristic frequencies. If the space is limited (as it always is in an instrument), the wave motion reflects from the fixed end of the guitar string or from the end of the tube in a wind instrument. The reflected wave then interacts with the original wave. Waves with the characteristic (*resonant*) frequencies of the instrument start to reinforce each other. This is called *constructive interference*. Where the two waves interfere constructively—with two crests combining or two troughs combining—you get an *antinode*. Where the two waves interfere destructively—a crest of one wave combining with a trough of the other wave—you get a *node*. At a node, the string or air does not move. At an antinode, the string or air vibrates the maximum amount. When the nodes

FIGURE 1.7: Standing Wave Apparatus

and antinodes stay in the same place, you have a *standing wave*.

In this experiment, the standing wave is caused by the small electric motor at the end of the string. The motor creates a transverse wave, which progresses to the point where the string is attached. You can change the number of nodes and antinodes by holding the string at different positions or, if you have a way to do it, by changing the speed of the motor.

The standing wave created does not appear to move forward or backward (or up or down). However, there is a wave moving: A wave is reflecting from the end, and an interference wave is formed. As the motor keeps running, you can see the formation of the standing wave.

Standing waves occur with pretty much all instruments, including percussion, string, woodwind, and brass instruments.

SOUND PIPE

When air flows through the Sound Pipe (Figure 1.8), there is some disruption to the air flow. The disruption occurs because the pipe's surface is not smooth. The deviations from smooth air flow are called *turbulence*, and this turbulence creates regions of lower and higher pressure. These pressure waves are reflected from the ends of the pipe, and the waves can be heard as a tone when the wavelength of the reflected wave is the same as the wavelength of the incoming wave. In this case, the waves reinforce each other.

Sound waves can be formed inside the pipe because of the turbulence. Turbulence makes waves of many frequencies—or wavelengths. The frequency or wavelength that gets reinforced depends on the length of the pipe. The wave will intensify (due to constructive interference) if the wavelength is twice the length of the pipe, the length of the pipe, two-thirds the length of the pipe, and so on, as shown in Table 1.1. These waves, which get more intense (or louder), are called harmonics. The different harmonics in the pipe have different numbers of nodes and antinodes. For sound resonating in a tube, the lowest resonant

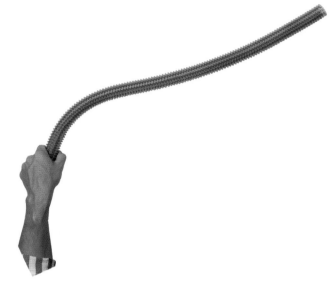

FIGURE 1.8: Sound Pipe

frequency is called the *fundamental frequency*, or the *first harmonic* (or $N = 1$). As you rotate the pipe faster and faster, you can hear the higher harmonics (second harmonic, third harmonic, and so on) of the fundamental frequency, but you won't hear the fundamental frequency itself because when you twirl the pipe slowly enough for the fundamental, there isn't enough air flow to make an audible sound. See Table 1.1 and Figure 1.9 for more details about these waves.

TABLE 1.1: HARMONICS IN A PIPE OPEN AT BOTH ENDS

Harmonic # (N)	# of wavelengths in the pipe	# of nodes	# of antinodes	Wavelength (in terms of the pipe wavelength, L)
1	1/2	1	2	Wavelength = 2L/1
2	1 or 2/2	2	3	Wavelength = 2L/2
3	3/2	3	4	Wavelength = 2L/3
4	2 or 4/2	4	5	Wavelength = 2L/4
5	5/2	5	6	Wavelength = 2L/5

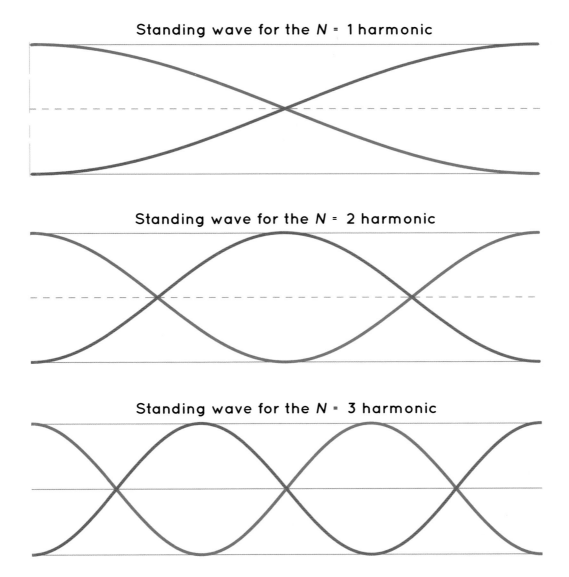

Standing wave for the *N* = 1 harmonic

Standing wave for the *N* = 2 harmonic

Standing wave for the *N* = 3 harmonic

FIGURE 1.9: Standing waves in an open pipe.

MUSIC BOX

The Music Box (Figure 1.10) illustrates how any system that oscillates or vibrates has a natural frequency of oscillation or vibration. External forces can make the system oscillate at any frequency, but if the external force acts on the system at its natural frequency, the amplitude of the oscillations will grow. The system is then in *resonance* with the external force.

The music box produces sound when the thin teeth of a metal comb vibrate. If the music box is played on a soft surface or in the air, the sound is not very loud. When you put the music box on a table or other hard surface, the sound gets much louder.

FIGURE 1.10 : Music Box

When you play the music box, the vibrating metal teeth move molecules of air, creating sound waves. When the vibrations move from the box to the surface of the table (through the base of the box), the surface of the table also starts to vibrate. The wider area now vibrating makes many more air molecules vibrate, and so the sound gets louder.

As you have seen, resonance is a phenomenon in which the vibration of one object gets another object to vibrate with the same frequency. Another way that this phenomenon can be studied is with two identical tuning forks. Tapping one tuning fork will cause the other one to start vibrating too.

Resonance is widely used in scientific research and medicine. For example, nuclear magnetic resonance spectroscopy uses the phenomenon of resonance. Nuclear magnetic resonance uses the magnetic properties of atomic nuclei to determine accurate physical and chemical properties of atoms and molecules.

If you did the second exploration, "Standing Wave," you already know that resonance is also an important phenomenon in music. For example, as a guitar string vibrates, the air molecules around the string start to vibrate, but more importantly, the vibrating strings make the wood of the (acoustic) guitar vibrate. The vibrating wood also makes air molecules vibrate—and far more air molecules vibrate due to the vibrating wood than from the vibrating strings. The vibrations of air molecules are sound waves, which can transfer energy.

FIGURE 1.11: The Doppler effect

DOPPLER BALL

If the source of the sound and the observer are stationary with respect to each other, the pitch or frequency of the sound that is heard remains the same. If the observer and the source of the sound either move further away from or get closer to one another, the sound changes. This phenomenon is called the *Doppler effect* (Figure 1.11). Sound moves through the air as a longitudinal or compression wave. When the source of the sound and the observer get closer to one another, the observer hears the sound higher in pitch; thus, the frequency of the sound is higher. The observer is then encountering crests of the waves more often. On the other hand, if the source and the observer move farther away from each other, the sound heard is lower because the observer now encounters crests of the waves less frequently.

This can be demonstrated with a wintry example: Two friends, Andy and Mandy, are having a snowball fight. Andy throws one snowball at Mandy every second from inside a snow fortress. After getting fed up with being hit by the snowballs at one-second intervals, Mandy starts running toward Andy. Now the balls seem to hit Mandy more often than once every second. From Mandy's

perspective, it seems as if Andy is now throwing the balls at her more frequently than before.

Finally Mandy starts running away from Andy's snow fortress. Now she concludes that Andy is throwing the snowballs at her less frequently because the hits are spaced farther apart from each other in time. According to Mandy, the frequency of the throws is lower.

The Doppler effect also explains the perception of *redshift* in astronomy. By allowing the light from a star to pass through a prism (or something similar), the light breaks up into a sequence of colors or wavelengths (Figure 1.12). That sequence of colors

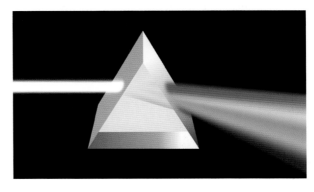

FIGURE 1.12: A prism separates white light into its component colors.

is called a spectrum, and when scientists closely examine the spectrum from a star, they see dark lines in it, like those shown in Figure 1.13.

FIGURE 1.13: The absorption spectrum from a star.

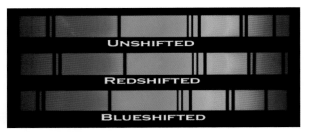

FIGURE 1.14: The Doppler shift of absorption lines.

The dark lines in this *absorption spectrum* are places where gases in the star have absorbed certain colors. If the star is moving toward or away from us, all the wavelengths in the spectrum will become shorter or longer, respectively—just like the sound waves did when the Doppler Ball was moving. The result is that the dark lines in the spectrum get shifted toward one end of the spectrum or the other (Figure 1.14).

By observing the redshift in a star's spectrum, the speed at which astronomical objects are moving toward or away from us can be determined. In a nutshell, the redshift phenomenon is similar to sound waves, only here it is light waves instead of sound waves. When the object moves away from the observer at a high speed, the frequency of the light observed is lower than the original frequency. Because frequency and wavelength are inversely proportional, the wavelength has increased. Now the lines in the spectrum are seen to have shifted toward longer wavelengths.

Web Resources

See how reflected waves result in a standing wave. Vary the wavelength.
www.colorado.edu/physics/2000/microwaves/standing_wave2.html

Explore standing waves on a string. Vary the frequency, amplitude, and length of the string.
http://ngsir.netfirms.com/englishhtm/StatWave.htm

See both standing waves and traveling waves, both transverse waves and longitudinal waves. Select the harmonic number. See where the nodes and antinodes are.
http://ngsir.netfirms.com/englishhtm/TwaveStatA.htm

Specify the length of a tube (open or closed) and see the resonant frequencies. See how they change with changes in air temperature.
http://hyperphysics.phy-astr.gsu.edu/hbase/waves/opecol.html

Make waves on a string. To make standing waves, click on "Oscillate" and set the damping to zero. Adjust the frequency and the string tension.
http://phet.colorado.edu/sims/wave-on-a-string/wave-on-a-string_en.html

See in slow motion how waves traveling in opposite directions can constructively interfere and create a standing wave. Click the ">" (play) button on the far right to start.
www2.biglobe.ne.jp/~norimari/science/JavaEd/e-wave4.html

Animation of resonance in a tube.
www.sciencejoywagon.com/physicszone/otherpub/wfendt/stlwaves.htm

Explanation of resonance in tubes (both open and closed).
http://dev.physicslab.org/Document.aspx?doctype=3&filename=WavesSound_ResonancePipes.xml

All about standing waves, nodes, and antinodes.
www.physicsclassroom.com/Class/waves/u10l4c.cfm

Visualization of the Doppler effect.
http://lectureonline.cl.msu.edu/~mmp/applist/doppler/d.htm

Relevant Standards

Note: The Next Generation Science Standards *can be viewed online at* www.nextgenscience.org/next-generation-science-standards.

CONNECTIONS TO NATURE OF SCIENCE

Science Models, Laws, Mechanisms, and Theories Explain Natural Phenomena

- Theories and laws provide explanations in science.

- Laws are statements or descriptions of the relationships among observable phenomena.

DISCIPLINARY CORE IDEAS

PS4.A: Wave Properties

- A simple wave has a repeating pattern with a specific wavelength, frequency, and amplitude. (MS PS4-1)

- A sound wave needs a medium through which it is transmitted. (MS-PS4-2)

CROSSCUTTING CONCEPTS

Energy and Matter

- Changes of energy and matter in a system can be described in terms of energy and matter flows into, out of, and within that system.

- Energy cannot be created or destroyed—only moves between one place and another place, between objects and/or fields, or between systems.

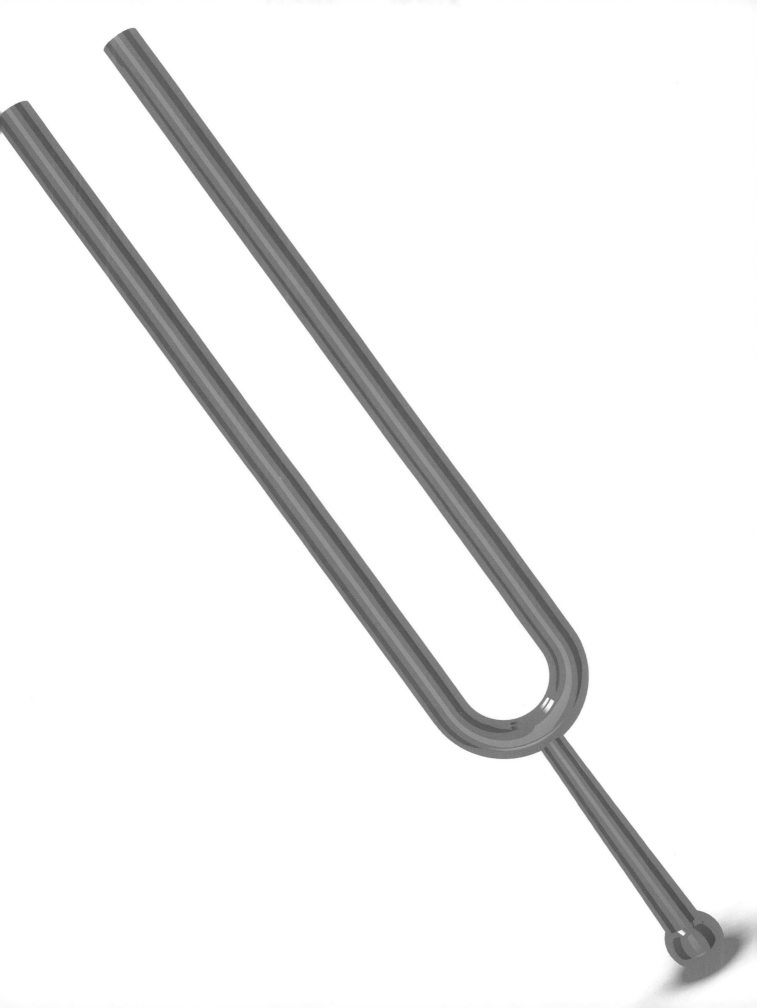

2

VISIBLE LIGHT AND COLORS

Color theory is the aspect of physics that can explain the colors on your computer screen or in your printer. Visible light can be split into different wavelength regions that are called colors. The color of an object is usually the color that the surface of that object is reflecting when white light shines on it. Some of the white light is absorbed and some is reflected or scattered from the surface. This usually determines what color we see.

In the eye there are *rod cells* and *cone cells*. The rod cells can detect the brightness of light but not its color; cone cells can detect color. There are three kinds of cone cells, and each is sensitive to certain ranges of wavelengths. One type of cone cell is sensitive to short wavelengths of light (such as blue), one type to medium wavelengths (such as green), and one type to long wavelengths (such as red). The sensitivity of the rod and cone cells will vary in different situations. The brain interprets the signals from the eyes and creates the sensation of different colors.

KEYWORDS
Knowing these terms will help you to enjoy the explorations.

prism

spectrum

refraction of light

reflection of light

RGB

CMYK

In this chapter, you will study the formation and sensation of colors. You will also learn about filters, primary colors, complementary colors, and adding and subtracting colors.

Exploration

FIGURE 2.1: Giant Prism

▼
SAFETY NOTE

Wear safety glasses or goggles.

PRISM PLAY

Sunlight is seen as white even though it consists of many colors. This can be seen with prisms, such as the Giant Prism (Figure 2.1) used in this exploration, or spectroscopes. In a prism, different wavelengths refract at different angles, making a spectrum in which you can see all the colors of a rainbow from violet to red. (The plural of *spectrum* is *spectra*.)

Procedure

1. Create a bright narrow beam of white light. Use a lightbulb or sunlight. You can use lenses and slits if they are available.

2. Dim the lights of the classroom.

3. Make the beam go through the prism so that a spectrum is formed.

Questions

- What colors can you observe in the spectrum?

- The red color refracts least. List the rest of the colors observed starting from red and ending with the color that refracts most.

- Find the wavelengths of different colors on the internet. List the colors starting from the shortest wavelength.

- Based on your results, how does the wavelength affect the refraction?

- You probably remember seeing the same colors in a rainbow. Explain now why there are rainbows in the sky and how they are created.

SPECTROSCOPE

With a Spectroscope (Figure 2.2) you can study the spectra of different light sources and the wavelengths of colors.

Procedure

1. Point the spectroscope toward various bright sources of light (but not directly at the Sun), including a fluorescent lamp, a lightbulb, daylight coming through a window, an LED, and a flame.

2. Notice the differences among the various spectra.

FIGURE 2.2: Spectroscope

Note: LED stands for light emitting diode, an electronic device that emits light when electricity passes through it. LEDs are now widely used in flashlights, traffic lights, and electronic displays and are becoming more popular in household lighting fixtures because of their low energy use and long life.

▼
SAFETY NOTE

Never touch the light source. It may be hot and burn skin.

Questions

*   What is the main difference between the spectra of sunlight and a fluorescent lamp?

*   What are the colors you found in the spectrum of a fluorescent lamp?

*   How is the spectrum from an LED different from the spectrum from a candle?

*   How do scientists use spectra to learn about the source of light? Explain.

FIGURE 2.3: Primary Color Light Sticks

▼
SAFETY NOTE

Use caution when moving around the lab with dimmed lights. Make sure there are no trip, slip, or fall hazards on the floor.

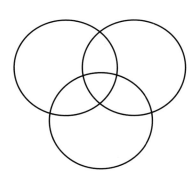

FIGURE 2.4: The outline for color combinations.

COLOR ADDITION AND SUBTRACTION

With the Primary Color Light Sticks (Figure 2.3) you can study the formation of colors and shadows.

Procedure

1. Dim the lights in the room.

2. Shine two of the lights, say the red and blue lights, onto a screen so that the colors overlap.

3. Repeat with other color combinations: red and green, blue and green. Take notes about what you observe.

4. Shine all three colored light sticks on one spot.

5. While the three colors are shining in one spot, make a shadow with your finger. Observe the color of the shadow as you move your finger to different areas.

Questions

- Red, green, and blue are called the primary colors of light. Why do you think these three colors are the primary ones?

- Draw a diagram of colors based on your work here. Use the overlapping circles in Figure 2.4, and label the different regions to show the primary colors and their combinations.

- What colors were the shadows when your finger was in front of the various lights? If you point all three color lights at one spot, what color will your finger's shadow be?

COMBINING COLORS

Next, we will look at the three primary colors and the formation of white light using the RGB Spinner and Snap Lights (Figure 2.5). We will also discuss color subtraction.

Procedure

1. Bend the snap light tubes (red, green, and blue) so that the thin capsules break inside the tube.

2. Attach the glowing snap lights to the spinner and dim the lights in the classroom.

3. Spin the device fast and see what color appears.

4. Take a piece of black tape and put it around the end of the red tube. Spin the tube again and notice the color.

5. Put a piece of tape close to the end of the green tube, but not as far as the tape on the red tube (see Figure 2.6). Spin the spinner and observe the new color.

6. Once again take some tape and cover a small area on the blue tube as shown in the picture.

Note: The RGB Snap Lights can be used repeatedly for up to 24 hours if they are stored in a cold place between uses.

Questions

- When first spinning, the combined color is close to white. What color is on the edge of the spinning circle when there is black tape on the red snap light?

- How about when the green color is blocked?

- When the blue color is blocked?

- What do the letters CMYK stand for? Do a web search to find out.

FIGURE 2.5: RGB Spinner and Snap Lights

▼
SAFETY NOTES

- Wear safety glasses or goggles.

- Use caution when moving around the lab with dimmed lights. Make sure there are no trip, slip, or fall hazards on the floor.

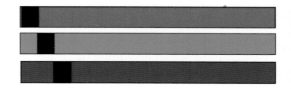

FIGURE 2.6: Tape placement for the three colored tubes.

PRISM PLAY

A beam of light refracts in the Giant Prism (Figure 2.7) and a spectrum is formed. Different materials refract light by different amounts. Every clear material has an *index of refraction*—a number that tells how much it refracts the light. The index of refraction is different for each color or each wavelength. The longest wavelength is refracted the least. This is why the color red refracts the least and violet the most. Water drops in the atmosphere can refract sunlight to create a spectrum, which can be seen as a rainbow.

FIGURE 2.7:
Giant Prism

SPECTROSCOPE

Spectroscopy is an area of physics that studies light, also called electromagnetic radiation. In a spectroscope (Figure 2.8), the light received passes through a narrow slit to a *diffraction grating*, which functions much like a prism. A diffraction grating has thousands of microscopic grooves that deflect the light at various angles, producing a sequence of colors ordered the same as in a rainbow.

FIGURE 2.8:
Quantitative
Spectroscope

When examining radiation emitted by the filament (incandescent) lamp, a continuous spectrum is observed: The spectrum shows all of the wavelengths of visible light from violet to red. Fluorescent lamps, LEDs, and many energy-saving bulbs produce a spectrum that shows only certain colors. The line spectrum is not continuous, but is seen only at certain wavelengths, looking like lines or bands of different colors.

The spectral lines produced by the fluorescent light can be explained by the formation of excited states of atoms. The electrons are transferred temporarily to a higher energy state, and the discharge of this excited state is observed as light. Each different chemical element has a different, unique set of spectral lines. This is why different kinds of lamps produce different spectra.

COLOR ADDITION AND SUBTRACTION

The primary colors are the colors that cannot be formed by mixing other colors. When combining different colors of light with the Primary Color Light Sticks (Figure 2.9), you are dealing with so-called *additive* color mixing. You get white light when the three primary colors are combined (in the proper amounts; see the tip in the "Combining Colors" analysis).

By combining the three primary colors (red, green, and blue) in pairs, you produce the subtractive primary colors yellow, magenta, and cyan. The yellow color is formed from red and green light. In yellow, there are usually no blue wavelengths of light at all. Similarly, the magenta color does not contain any green light wavelengths, and the cyan color does not have any red light wavelengths.

The colors produced by a computer monitor or TV screen are based on the RGB method—combining red (R), green (G), and blue (B) in different amounts. Each screen pixel is a single light source for either red, green, or blue light. The colors

FIGURE 2.9: Primary Color Light Sticks

COMBINING COLORS

The RGB Spinner and Snap Lights (Figure 2.11) allow you to explore subtractive colors. These colors (which work for paints, dyes, and inks) are cyan (C), magenta (M), yellow (Y), and black (K). A particular color of paint removes (by absorbing) some parts of the white light spectrum. Yellow paint absorbs blue light, but reflects red and green; magenta paint absorbs green, but reflects red and blue; and cyan paint absorbs red, but reflects green and blue. In printing, black provides a better black than could be achieved by combining cyan, magenta, and yellow. Black is symbolized by K, for "key," referring to the key plate used by printers. This plate provides additional detail in an image, usually in black ink.

Printing colors are usually abbreviated as CMYK, as mentioned above. Four-color printing means that these four colors are used for the formation of color images. When using the CMYK colors, say in an image-editing computer program, the intensities of the four CMYK colors are saved for each pixel.

Tip: The three colored snap lights probably won't have the correct relative brightnesses to make white when they are spun. If you see a color when spinning the lights, that tells you which color is too bright relative to the others. Put a thin strip of black tape along the tube that's too bright to block some of its light and make the colors more balanced. That way, you get as close to white as possible when you spin the spinner.

FIGURE 2.11: RGB Spinner and Snap Lights

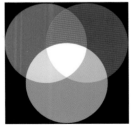

FIGURE 2.10: Color combinations

shown on the screen are formed by adjusting the intensities of these colors.

In the exploration, you also observed the colors of the shadows. When each primary color was blocked, you saw the differently colored shadows as yellow (when blue was blocked), cyan (when red was blocked), and magenta (when green was blocked) (Figure 2.10).

Web Resources

The electromagnetic spectrum and the visible spectrum.
http://csep10.phys.utk.edu/astr162/lect/light/spectrum.html

A PowerPoint presentation about color.
www.msb-science.com/color-ch27-E-Anthony.ppt

Color wheel and color vocabulary.
https://docs.google.com/document/d/1GyO_HrnKE-MvSTItB6F3dT6BdfS_nM1_C94pg-BqgE30/edit?hl=en

Additive color mixing applet.
http://lectureonline.cl.msu.edu/~mmp/applist/RGBColor/c.htm

Additive (light) and subtractive (paint) color mixing applet.
http://cs.brown.edu/exploratories/freeSoftware/repository/edu/brown/cs/exploratories/applets/combinedColorMixing/combined_color_mixing_java_browser.html

See the absorption or emission spectrum from any chemical element.
http://jersey.uoregon.edu/vlab/elements/Elements.html

See a continuous spectrum corresponding to any temperature. Notice that when you set it for the temperature of the Sun, the color is approximately white.
http://phet.colorado.edu/sims/blackbody-spectrum/blackbody-spectrum_en.html

Relevant Standards

Note: The Next Generation Science Standards *can be viewed online at* www.nextgenscience.org/next-generation-science-standards.

CONNECTIONS TO NATURE OF SCIENCE

Science Models, Laws, Mechanisms, and Theories Explain Natural Phenomena

Theories and laws provide explanations in science.

- Laws are statements or descriptions of the relationships among observable phenomena.

DISCIPLINARY CORE IDEAS

PS4.B: Electromagnetic Radiation

- When light shines on an object, it is reflected, absorbed, or transmitted through the object, depending on the object's material and the frequency (color) of the light. (MS-PS4-2)

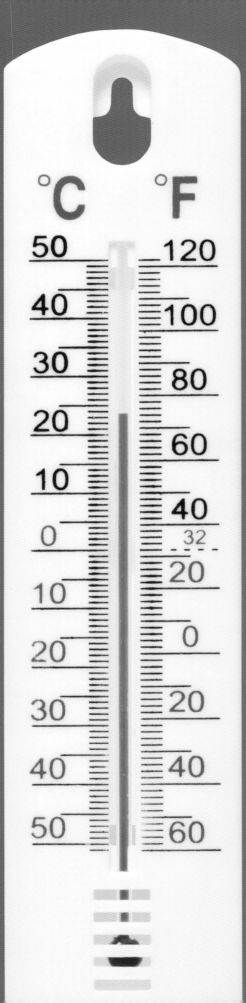

THERMODYNAMICS

Thermodynamics is a very important part of physical science. When discussing thermodynamics, we can explore heat, thermal expansion, and insulation, as well as changes in state, thermal equilibrium, and thermal energy.

Temperature tells us how hot or cold an object is. Temperature is also a measure of how quickly atoms and molecules vibrate. The faster the particles move, the warmer the object is. We can measure temperature with a thermometer. The number it reads tells us how warm or cold it is compared with certain reference points. In some temperature scales, the reference points are the freezing point of water and the boiling point of water. In the Celsius scale, the numbers are 0°C for freezing and 100°C for boiling. In the Fahrenheit scale the corresponding reference points are 32°F and 212°F, respectively. In the Kelvin scale, the lower reference point is absolute zero, which is approximately –273°C (or –460°F). In the Kelvin scale, there is no upper reference point, and one degree is the same amount of temperature difference as in the Celsius scale.

KEYWORDS
Knowing these terms will help you to enjoy the explorations.

thermometer

Celsius scale

Fahrenheit scale

energy

heat

An understanding of heat and thermodynamics will help you understand why heat behaves the way it does and how you can use this understanding for your benefit. For example, why is a frying pan made of metal while its handle is made of plastic or wood? Why does the fan on the table make you feel more comfortable on hot summer days, even though the fan doesn't actually make the air any cooler?

The experiments in this chapter illustrate some of the key thermal physics phenomena from different perspectives.

FIGURE 3.1: Radiation Cans

▼ SAFETY NOTES

- Wear safety glasses or goggles.
- Glass thermometers are fragile and can shatter. Handle with care.
- Do not use mercury thermometers.

RADIATION CANS

In this experiment, you will study how the color of an object influences its temperature. You will need three bottles (white, silver, and black), such as the Radiation Cans (Figure 3.1), water, and a thermometer for each bottle.

Procedure

1. Pour exactly the same amount (e.g., 100 ml) of the same temperature water into each bottle.

2. Put a thermometer into each bottle.

3. Place the bottles in direct sunlight or near a hot, bright light.

4. Observe and record the temperatures in the bottles. Graph your data to see how the temperature changed in the different bottles.

Questions

- How did the color of the bottle affect how much the water warmed up?

- Which bottle changed temperature fastest?

- Show on your graph where you can see the differences in warming.

- How is the information you collected useful on hot sunny days or cold winter days?

FIGURE 3.2: Ice Melting Blocks

MELTDOWN

With the Ice Melting Blocks (Figure 3.2), you will explore some thermal properties of different materials.

Procedure 1

1. Briefly touch the surfaces of the two blocks.

Questions

- Which block feels colder?

- Predict on which block an ice cube will melt faster. Explain your reasoning.

Procedure 2

1. Put the O-rings on the blocks, and then set the ice cubes on the blocks.

2. Observe the melting of the ice. Take notes.

3. After a couple of minutes of melting, touch the blocks again (where there's no ice or water) and sense the temperatures.

4. Feel how warm or cold the table is under the blocks. Take notes.

▼
SAFETY NOTES

- Wear safety glasses or goggles.

- Immediately wipe up any splashed water to prevent a slip or fall hazard.

Questions

- Why does one block feel much colder than the other one?

- How do you explain your observations of the temperature of the table under the blocks?

- Why did the ice cube on one block melt faster than on the other block?

FIGURE 3.3: Ball and Ring

▼
SAFETY NOTES

- Wear safety glasses or goggles.

- Use caution when working with Bunsen burner or other heat source. They can burn skin.

- Never leave a heat source unattended when it is hot.

- Do not handle the metal part of the Ball and Ring equipment when it is heated. It can burn skin.

BALL AND RING

Here you will investigate some effects of changes in temperature. In addition to the Ball and Ring (Figure 3.3), you will need a Bunsen burner or some other kind of heater.

Procedure

1. Check that the ball fits through the ring.

2. Predict what will happen when the ball is heated. Explain your reasoning.

3. Heat the ball and try again.

4. Cool the ball down in water before the next experiment.

5. Predict whether the ball will fit through the ring if the ring is heated up. Explain your reasoning.

6. Heat up the ring and check your prediction.

Questions

- First the ball was heated, and then the ring. Describe your observations in each part of your exploration, and suggest an explanation for the physics that was involved in each part.

- Can you think of ways in which this phenomenon could be useful?

- How might engineers take this phenomenon into account when they, for example, design a bridge?

DRINKING BIRD

The Drinking Bird (Figure 3.4) demonstrates several phenomena related to thermodynamics. The demonstration is ambitious— the physics can be quite a challenge—but very interesting.

In addition to the bird, you will need a glass of water.

Procedure

1. Put some water in a glass and dunk the Drinking Bird's head in the water. Add more water so that the glass is completely full.

2. Set the drinking bird down so that when it tips over its beak goes in the water.

Questions

- What makes the bird's balance change?

- How does evaporation affect the bird's pecking?

- Does pressure play any role here? Why?

- Does anything different happen if you warm up the lower liquid container with your hand?

- What causes the pecking to begin?

- Is it a perpetual motion machine? Explain.

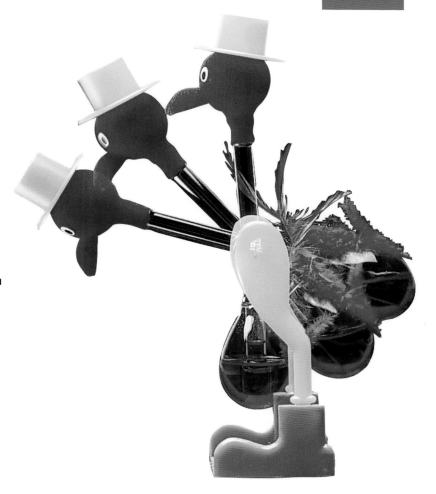

FIGURE 3.4: Drinking Bird

▼
SAFETY NOTE
Wear safety glasses or goggles.

FIGURE 3.5:
Fire Syringe

FIRE SYRINGE

With the Fire Syringe (Figure 3.5) you can explore the connection between pressure and temperature. In addition to the Fire Syringe itself, you will need a tiny bit of cotton.

▼

SAFETY NOTE

Wear safety glasses or goggles.

Procedure

1. Drop a tiny piece of cotton (about the size of a pencil point) into the cylinder. If it doesn't fall to the bottom, push it down with a thin rod.

2. Screw the piston into place with it all the way up.

3. Turn off the lights in the room.

4. Sharply—as fast as possible—press the piston down.

Questions

- What happened in the cylinder?

- Explain your observations from a physics point of view.

- Where could this phenomenon be useful?

Analysis

RADIATION CANS

Note on terminology: In physics, *heat* is energy moving from a hotter object to a cooler object. That is somewhat different from the popular use of the word *heat* to refer to the *thermal energy* contained in an object.

Thermal energy can be transferred from one object to another by *conduction* if the two objects are touching each other or by radiation, such as when infrared light energy from the wires in a toaster is absorbed by the bread to create toast. The Radiation Cans (Figure 3.6) explore the latter phenomenon.

If thermal energy is transferred to a fluid (liquid or gas), then the energy can be carried away with the flow of the fluid. Transporting of thermal energy by a fluid is called *convection*. An object gains thermal energy by conduction or by absorbing radiation. When thermal energy flows directly between two materials, it always flows from the hotter material to the colder material. That flowing energy is heat. When the temperatures of the objects in a system are equal, we say that the system is in *thermodynamic equilibrium*.

After cool water is poured in the bottles, the black bottle warms up the fastest. The black surface absorbs all wavelengths of visible light most efficiently. By absorbing that radiation, the bottle heats up. The white bottle and the silver bottle reflect some of the wavelengths of light that strike the bottles, so the change in temperature is less than in the case of the black bottle. This is why wearing dark clothes keeps you warmer on a cold sunny day. On a hot, sunny day, you might want to wear clothing of lighter colors to reflect more of the Sun's radiation and stay cooler.

FIGURE 3.6: Radiation Cans

FIGURE 3.7: Ice Melting Blocks

MELTDOWN

Thermal energy is the random movement or vibration of particles in an object. The faster the particles in an object vibrate, the hotter the object is. When a warmer object (like one of the Ice Melting Blocks [Figure 3.7]) touches a cooler object (like the ice cube), the faster moving particles in the warmer object hit the slower moving particles in the cooler object. This slows down the particles in the warmer object, making it cooler, and it speeds up the particles in the cooler object, making it warmer. In this way, thermal energy is transferred from the warmer object to the cooler object by a process called *thermal conduction*.

Some materials are better heat conductors than others. Materials that do not conduct heat well are called *thermal insulators*. Air is a thermal insulator. However, for air to work well as an insulator, it must be confined in small spaces so that it can't flow and transport heat away via convection. In a block of something like Styrofoam, the air is confined in very small spaces so that it can't flow. That makes a foam block—such as one of the two blocks in this experiment—a good insulator.

The second block is made of aluminum. In metals there are electrons that greatly increase the thermal conduction. These electrons move freely among the metal molecules. The aluminum block feels cold because it is a good thermal conductor, and thermal energy is rapidly transferred from your hand to the block. In the insulator there is not that kind of conduction present so your hand doesn't lose thermal energy very fast, and the block does not feel cold in your hand. In other words, even though both blocks start out at the same temperature (room temperature), the aluminum block *feels* colder because it conducts heat out of your hand more rapidly.

The ice cube put on the aluminum block melts faster because aluminum is a good conductor and transfers thermal energy into the ice cube quite effectively. The energy is transferred as long as there is a difference in temperature between the block and the ice cube. On the insulator, we cannot see the cube melting down as fast because there is not that much thermal energy being transferred from the block to the cube.

NATIONAL SCIENCE TEACHERS ASSOCIATION

FIGURE 3.8:
Ball and Ring

BALL AND RING

If the air around the ice cube isn't moving much, there will be a cold, insulating layer of air that prevents a lot of thermal energy from flowing from the air to the ice cube. The cube would melt faster if air was blowing past it.

To melt ice requires energy from the surroundings, so the conductor (aluminum) block feels—and really is—cooler right after the experiment. The table surface under the conductor block feels colder too. As the aluminum block became cooler (having lost thermal energy to the ice), energy flowed from the table into the aluminum block, making the table cooler as well. Under the insulator block, the temperature of the table hasn't changed much.

Tip: If there are not enough blocks for each group to try the experiment, do this as a demonstration with a document camera. If you want a more precise measure than simply feeling the blocks can provide, you can attach thermal sensors to the blocks.

When the metal ball of the Ball and Ring (Figure 3.8) is heated up, the thermal motion (the vibration of atoms and molecules) increases, causing the particles to take up more space. After the ball expands, it will not fit through the ring.

When the ball has cooled down, you can see what happens when the ring is heated up. You may have thought that the ball would not fit through the ring because the thermal expansion acts in all directions, including pushing in the inner edge of the ring, making the hole smaller. However, the proportions of the ring remain the same when the ring is heated up. This situation can be compared with enlarging a photo. If you enlarge a photo of a ring, both the outer and inner edges of the ring increase in size. Likewise, the hole in the ring expands when the ring is heated up.

Another way to look at it is that when the expansion occurs, the distance between any two points on the ring will increase. This is true even if those two points are on opposite sides of the hole. Thus the diameter of the hole increases.

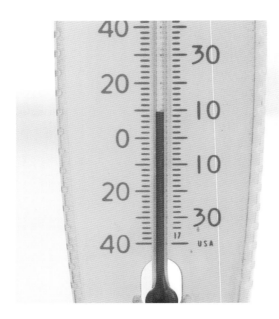

FIGURE 3.9: Thermometer

Thermal expansion is also the explanation for how many thermometers work. If it's the kind of thermometer with liquid in a tube (Figure 3.9), when it gets warmer the liquid in the tube (and in the bulb at the bottom) expands, and the liquid rises in the tube.

The rising liquid in the thermometer therefore provides a measure of the increasing thermal energy.

Thermal expansion is an everyday phenomenon that can not only be used to explain the function of a thermometer or the change in size of metallic materials at different temperatures, but that must also be

FIGURE 3.10: Bridge expansion joint

taken in account when building houses or bridges. Bridges are made with expansion joints, which have extra room for the expansion that occurs at higher temperatures. Expansion joints like the one in the Auckland Harbour Bridge (see Figure 3.10) in New Zealand allow bridges to expand due to heat without buckling.

DRINKING BIRD

The Drinking Bird (Figure 3.11) consists of two glass bulbs connected by a thin tube. One bulb is the bird's head, and the other bulb is in the bird's lower body. At first the pressure and temperature are the same in both bulbs, all the liquid is at the bottom, and the bird remains in balance in an upright position. Although all the liquid starts out at the bottom, there is vapor from that liquid in the head.

After the head (or beak) is soaked, water starts to evaporate from the head. The evaporation cools down the head and the vapor inside it. When the

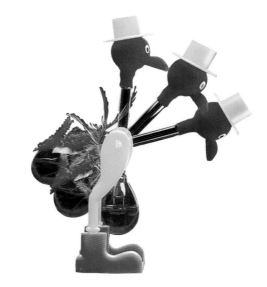

FIGURE 3.11: Drinking Bird

temperature goes down the pressure drops too, and some of the vapor condenses. Now the pressure is lower inside the head than in the lower bulb. Because of that pressure difference, the liquid in the lower bulb gets pushed up through the tube. The rising liquid changes the balance of the bird, and it begins to tip over. When the liquid reaches the head and starts to fill it, the bird is so top-heavy that it completely tips over, and the beak goes into the water again. When the bird tips over, the pressures in the two bulbs can equalize, the liquid flows back down, and the bird swings back to its starting position. Then the same process starts again.

You can also make the liquid flow up the tube by warming the lower bulb with your hand. The rise in temperature causes increased vaporization of the liquid and increased pressure of the gas in the lower container, causing the liquid to move up the tube.

This relation between temperature and pressure is described by Gay-Lussac's law, which expresses the fact that the ratio between pressure and temperature remains constant in a fixed volume. Thus, as the temperature rises, the pressure rises proportionately.

FIGURE 3.12: Fire Syringe

FIRE SYRINGE

When you push the piston of the Fire Syringe (Figure 3.12) down, the volume of the air inside the cylinder quickly decreases. Consequently, the speed and frequency of collisions of the air particles increases. In addition, the piston moving and colliding with the particles makes the particles speed up. Both the kinetic energy of the particles and the temperature increase in the cylinder. The temperature rises to the ignition temperature of cotton (350–400°C), and the cotton catches on fire. This process, in which the gas heats up so fast that it doesn't have time to lose much heat to the environment, is called an *adiabatic* process.

Adiabatic processes are used, for example, in diesel engines. The increase in pressure inside the cylinder heats up the diesel fuel, which then catches fire. In gasoline engines, the fuel is ignited by a spark plug.

Web Resources

Experiment with pressure, volume, and temperature changes with a piston.
www.mhhe.com/physsci/physical/giambattista/thermo/thermodynamics.html

Physics Applets on a variety of topics, including thermodynamics.
http://jersey.uoregon.edu/

Compare heat conduction through two materials of different thermal conductivities.
http://energy.concord.org/energy2d/thermal-conductivity.html

Explore how energy flows and changes through a system.
http://phet.colorado.edu/en/simulation/energy-forms-and-changes

Learn how the properties of gas (volume, heat, and so on) vary in relation to each other.
http://phet.colorado.edu/en/simulation/gas-properties

A Boyle's law experiment.
www.chm.davidson.edu/vce/gaslaws/boyleslaw.html

A Boyle's law animation.
www.grc.nasa.gov/WWW/k-12/airplane/aboyle.html

A Boyle's law worksheet.
www.grc.nasa.gov/WWW/k-12/BGP/Sheri_Z/boyleslaw_act.htm

A Charles's law experiment.
www.chm.davidson.edu/vce/gaslaws/charleslaw.html

Some animated activities and worksheets on gas laws.
www.nclark.net/GasLaws

A Charles and Gay-Lussac's law activity.
www.grc.nasa.gov/WWW/k-12/airplane/glussac.html

A Charles and Gay-Lussac's law animation.
www.grc.nasa.gov/WWW/k-12/airplane/aglussac.html

Relevant Standards

Note: The Next Generation Science Standards *can be viewed online at* www.nextgenscience.org/next-generation-science-standards.

PERFORMANCE EXPECTATIONS

MS-PS3-3

Apply scientific principles to design, construct, and test a device that either minimizes or maximizes thermal energy transfer.* [Clarification Statement: Examples of devices could include an insulated box, a solar cooker, and a Styrofoam cup.] [Assessment Boundary: Assessment does not include calculating the total amount of thermal energy transferred.]

MS-PS3-4

Plan an investigation to determine the relationships among the energy transferred, the type of matter, the mass, and the change in the average kinetic energy of the particles as measured by the temperature of the sample. [Clarification Statement: Examples of experiments could include comparing final water temperatures after different masses of ice melted in the same volume of water with the same initial temperature, the temperature change of samples of different materials with the same mass as they cool or heat in the environment, or the same material with different masses when a specific amount of energy is added.] [Assessment Boundary: Assessment does not include calculating the total amount of thermal energy transferred.]

MS-PS3-5

Construct, use, and present arguments to support the claim that when the kinetic energy of an object changes, energy is transferred to or from the object. [Clarification Statement: Examples of empirical evidence used in arguments could include an inventory or other representation of the energy before and after the transfer in the form of temperature changes or motion of object.] [Assessment Boundary: Assessment does not include calculations of energy.]

SCIENCE AND ENGINEERING PRACTICES

Developing and Using Models

Modeling in 6–8 builds on K–5 and progresses to developing, using and revising models to describe, test, and predict more abstract phenomena and design systems.

- Develop a model to predict and/or describe phenomena. (MS-PS1-1),(MS-PS1-4)

- Develop a model to describe unobservable mechanisms. (MS-PS1-5)

Analyzing and Interpreting Data

Analyzing data in 6–8 builds on K–5 and progresses to extending quantitative analysis to investigations, distinguishing between correlation and causation, and basic statistical techniques of data and error analysis.

- Analyze and interpret data to determine similarities and differences in findings. (MS-PS1-2)

Obtaining, Evaluating, and Communicating Information

Obtaining, evaluating, and communicating information in 6–8 builds on K–5 and progresses to evaluating the merit and validity of ideas and methods.

- Gather, read, and synthesize information from multiple appropriate sources and assess the credibility, accuracy, and possible bias of each publication and methods used, and describe how they are supported or not supported by evidence. (MS-PS1-3)

CONNECTIONS TO NATURE OF SCIENCE

Science Models, Laws, Mechanisms, and Theories Explain Natural Phenomena

- Theories and laws provide explanations in science.

- Laws are statements or descriptions of the relationships among observable phenomena.

DISCIPLINARY CORE IDEAS

PS3.A: Definitions of Energy

- The term "heat" as used in everyday language refers both to thermal energy (the motion of atoms or molecules within a substance) and the transfer of that thermal energy from one object to another. In science, heat is used only for this second meaning; it refers to the energy transferred due to the temperature difference between two objects. (secondary to MS-PS1-4)

- The temperature of a system is proportional to the average internal kinetic energy and potential energy per atom or molecule (whichever is the appropriate building block for the system's material). The details of that relationship depend on the type of atom or molecule and the interactions among the atoms in the material. Temperature is not a direct measure of a system's total thermal energy. The total thermal energy (sometimes called the total internal energy) of a system depends jointly on the temperature, the total number of atoms in the system, and the state of the material. (secondary to MS-PS1-4)

PS3.A: Definitions of Energy

- Temperature is a measure of the average kinetic energy of particles of matter. The relationship between the temperature and the total energy of a system depends on the types, states, and amounts of matter present. (MS-PS3-3) (MS-PS3-4)

PS3.B: Conservation of Energy and Energy Transfer

- The amount of energy transfer needed to change the temperature of a matter sample by a given amount depends on the nature of the matter, the size of the sample, and the environment. (MS -PS3-4)

- Energy is spontaneously transferred out of hotter regions or objects and into colder ones. (MS-PS3-3)

CROSSCUTTING CONCEPTS

Energy and Matter

- Energy may take different forms (e.g., energy in fields, thermal energy, energy of motion). (MS-PS3-5)

- The transfer of energy can be tracked as energy flows through a designed or natural system. (MS-PS1-6)

Patterns

- Different patterns may be observed at each of the scales at which a system is studied and can provide evidence for causality in explanations of phenomena.

Cause and Effect

- Empirical evidence is required to differentiate between cause and correlation and make claims about specific causes and effects.

- Systems can be designed to cause a desired effect.

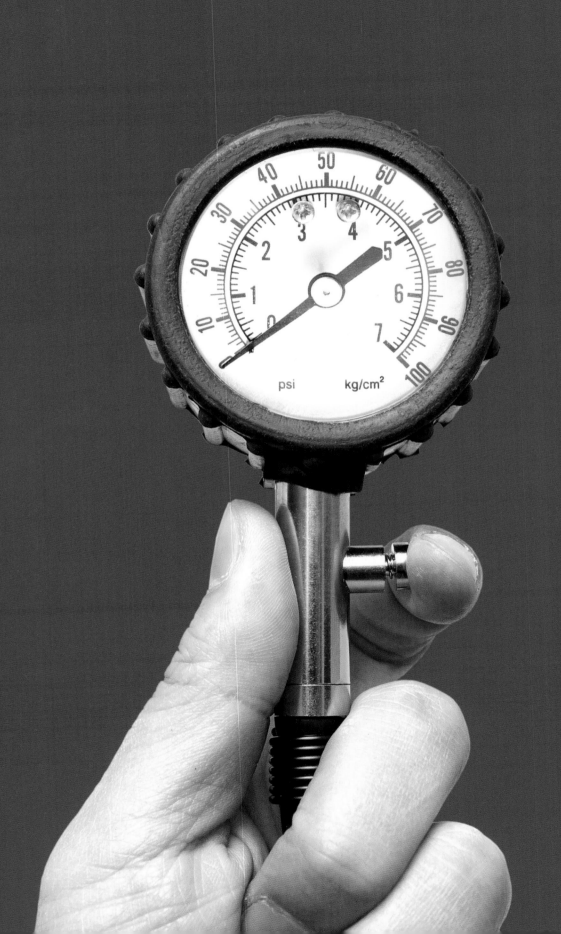

4

AIR PRESSURE

Air pressure has a significant role in everyday life. Pressure-related concepts such as vacuum, excess pressure, and high and low pressure are encountered almost daily. We talk about negative pressure, vacuum pumps, atmospheric pressure, and pressure chambers. Bicycle tires can be inflated to a pressure that is several times higher than the atmospheric pressure. Passengers in airplanes feel pressure changes during the flight. Air pressure also significantly affects the weather.

KEYWORDS
Knowing these terms will help you to enjoy the explorations.

atmosphere

air pressure

vacuum

force

Air pressure depends on the temperature and the number of gas molecules in a certain volume. As the temperature gets warmer, the gas molecules move faster and collide with surfaces more often. This makes the pressure higher. Reducing the volume of a container makes the pressure increase, and decreasing the temperature or increasing the volume makes the pressure decrease.

In the *International System of Units* (SI units), the unit of pressure is *newtons* per square meter (N/m^2). This unit is also known as the *pascal* (Pa): when the force of one newton is applied to one square meter of area, the pressure is equal to one pascal.

The weight of the atmosphere above an object pushes on the object with a certain amount of air pressure. However, air pressure does not only push downward, but in all directions. So there is air pressure not only on the floor of your classroom, but also on the walls and ceiling. Near the surface of the Earth, the air pressure averages 101.3 kPa (1 kPa = 1,000 Pa), and it decreases at higher altitudes. The other often-used unit of pressure is the *bar*. It corresponds approximately to the normal atmospheric pressure. One bar is defined as 100 kPa.

In this chapter, you will learn about concepts and phenomena concerning air pressure.

Exploration

FIGURE 4.1: Atmospheric Pressure Mat

IT'S A HOLD-UP!

With the Atmospheric Pressure Mat (Figure 4.1), you will study the force caused by a pressure difference as well as the reason for the pressure difference.

Procedure

1. Choose an object to lift. You can lift objects that weigh up to 20 kg (44 lb) with the mat, so you can choose, for instance, a chair or a small table as long as it has a smooth surface.

2. Put the mat on the object, making sure that the surface is both smooth and clean, and lift from the hook attached to the mat. Be careful about the balance of the object as you lift it—the object can accidentally detach and fall.

3. Repeat the procedure by lifting various (for example, lighter or heavier) objects with different surfaces.

4. Put something like a towel or a tablecloth between the mat and the object you are lifting or try to lift the cloth itself.

Questions

• What happened to the mat when it attached to the surface and you lifted it with the hook?

• What makes it possible to lift objects with the Atmospheric Pressure Mat?

• Based on your observations, on what kind of surfaces will lifting with the mat not work? Why is that?

• How could you improve the Atmospheric Pressure Mat to lift even heavier loads?

Pressure and force have a relation:

$$P = \frac{F}{A}$$

where P is the pressure, F is the force, and A is the area over which the force is applied. Estimate or calculate how great a load could be lifted if there was a perfect vacuum under the mat.

FIGURE 4.2: Suction cups

PRESSURE POWER

You can study partial vacuums and pressure differences with suction cups (Figure 4.2).

Procedure

1. Compress the cups against each other and then pull them apart.

2. Repeat the experiment with some paper or cloth between the cups.

3. Invent something new for which these cups can be useful.

Questions

- You may hear a sound when compressing the cups together. Why?

- What causes the force that keeps the cups together?

- How can you more easily separate the cups?

- What are suction cups normally used for?

▼
SAFETY NOTES

- Wear safety glasses or goggles.

- Make sure your arms are in an area free of any objects to prevent injury when pulling the suction cups apart.

PRESSURE GLOBE

The Pressure Globe (Figure 4.3) is used when studying the concepts of forces produced by pressure differences or a partial vacuum. The Pressure Globe has a balloon inside it and a plug on the bottom.

Procedure

1. Place the plug in the bottom of the globe and try to inflate the balloon.

2. Remove the plug. Blow up the balloon, plug the globe when the balloon is filled, and then take your mouth off the balloon.

3. Remove the plug again.

4. Blow up the balloon again, and plug the globe when the balloon is filled. Then, put 100 ml of water into the balloon. Over a sink, or standing outside, remove the plug.

Questions

• Were you able to inflate the balloon when the plug was in the bottom? Explain why or why not.

• What happens when you inflate the balloon while the plug is removed? Why?

• Explain what happens when you remove the plug while the balloon is inflated.

• What happened when the water was in the balloon and you removed the plug? Why?

FIGURE 4.3: Pressure Globe

▼
SAFETY NOTES

• Wear safety glasses or goggles.

• Immediately wipe up any splashed water to prevent a slip or fall hazard.

WATER ROCKET

The water rocket is a very impressive experiment, which can be used to learn about increased pressure or pressure differences. When done carefully, the experiment is safe, but we suggest that you use the Bottle Rocket Launcher (Figure 4.4) with the help of an assistant. You should use a pump that has a pressure gauge.

Procedure

1. Read all of the instructions that came with the water rocket.

2. Set up the launch pad with the stake securely in the ground, and make sure that there are no people, cars, buildings, power lines, trees, and so on close to your launch station.

3. Fill the bottle half full with water.

4. Attach the bottle to the launch pad's plug, and put the cotter pin in place.

5. Extend the other end of the launch cord as far as possible.

6. Attach the pump hose to the valve stem on the end of the pressurizing hose.

7. Check for stability by pulling on the string to make sure the stand is well anchored and won't tip over when launching.

8. Start pumping. Monitor the rocket at all times in case something starts to go wrong.

9. When there are approximately five bars of pressure in the bottle, do the countdown and launch the rocket by sharply pulling the cotter pin away.

Questions

• Predict how high the rocket will go.

• Why does the rocket leave the ground?

• Why did you need to have water in the bottle?

• What do you see in the bottle after the rocket has landed? Why?

FIGURE 4.4: Bottle Rocket Launcher

▼
SAFETY NOTES

• Wear safety glasses or goggles.

• Never stand over or near the water rocket while it is being pressurized or launched.

• Stay clear of the stake—it presents a potential impalement hazard.

• Only perform this experiment outside in an open field, never inside.

IT'S A HOLD-UP!

The Atmospheric Pressure Mat (Figure 4.5) allows you to explore the relationship between pressure and volume in gas. The gas pressure in a container is caused by molecules colliding with the walls of the container. As you pull up on the mat for the exploration, the volume under the mat starts to expand. However, the (small) amount of air under the mat remains the same and simply spreads out more in that larger volume, thus reducing the density of the air. Then, the collisions of molecules with the table become less frequent, meaning that there is less air pressure pushing down on the top of the table.

In contrast, there is much more air pressure pushing up on the bottom of the table. The difference between these two pressures—on the top of the table and the bottom of the table—provides enough force to hold the table up. In other words, the upward air pressure on the bottom of the table is greater than the downward pressure on the top of the table by

an amount equal to the weight of the table: The table remains suspended.

The surface of the object being lifted must be smooth so that air molecules do not leak in under the mat. A rough surface such as a tablecloth lets air leak in, thus equalizing the pressure above and below the table. Then gravity makes the table fall.

When calculating the maximum load that can be lifted, you might assume that there is a perfect vacuum under the mat. In real life, you cannot have a perfect vacuum, and the mat will not hold as much weight as you calculate. Cleaning the surface and the mat as well as using a larger mat will increase the ability to lift heavier loads.

PRESSURE POWER

Pressing the suction cups (Figure 4.6) together squeezes the air out from between them. The escaping air makes the edges of the cups vibrate, producing a sound that you can hear.

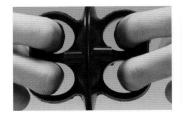

FIGURE 4.6: Suction cups

When you start to pull the cups apart, the volume between them increases, but the amount of air molecules remains the same. The pressure therefore becomes lower as a result of less frequent collisions of the air molecules against the inner surfaces of the cups. The air pressure outside the cups is still the same, so it is pushing the cups together. Suction cups should really be called pressure cups (and the official toy name is Atmospheric Pressure Cups) because it is pressure that holds them together. Suction (or pressure) cups are widely used. For example, you can attach a navigation device to the windshield of a car, and they can also be helpful when handling glass.

FIGURE 4.5: Atmospheric Pressure Mat

FIGURE 4.7: Pressure Globe

PRESSURE GLOBE

The explorations with the Pressure Globe (Figure 4.7) work because of the differences in pressure between the air in a balloon and the air in a hollow glass ball. With the plug in the globe, blowing air into the balloon is difficult because the air pressure in the globe (and a small amount of elasticity from the balloon) keep the balloon compressed. While blowing, you compress the air between the balloon and globe, and the increased pressure pushes in on the balloon even more. You can blow a small amount of air into the balloon, but you quickly reach a point at which you cannot blow any more in.

Once you remove the plug and blow into the balloon, you can make the pressure inside the balloon higher than the outside air pressure and expand the balloon. When you stop blowing air into the balloon, it deflates because of its elasticity increasing the pressure enough to force the air out.

If you plug the globe after blowing up the balloon, the balloon stays inflated. The balloon tries to contract, but there is no path for the air to get behind the balloon. A slightly lower pressure is created between the balloon and the globe.

Finally, in the step with water, the water stays in the balloon when the globe is plugged because the reduced air pressure in the globe is not enough to push the water out. When the plug is removed, air flows into the globe and the pressure increases to the point where the pressure is the same on both sides of the balloon. At that point, the elasticity of the balloon pushes the water out.

WATER ROCKET

The Bottle Rocket Launcher (Figure 4.8) provides one of the most dramatic explorations. Pumping air into the water rocket increases the pressure inside the bottle, which is higher than the surrounding, normal air pressure. When the rocket is launched, the higher inside pressure forces the water out as the pressure inside returns to the nor-

FIGURE 4.8: Bottle Rocket Launcher

mal air pressure. The water is moving downward so (because momentum is conserved) the rocket starts to move upward. Another way of looking at it is that the pressure inside the bottle pushes up on the top of the bottle. The high speed of the rocket is due to the small mass of the rocket compared to the mass of water ejected, and secondly because the water is moving so fast.

After the landing, you can see a small cloud inside the bottle. When the moist air in the bottle suddenly expanded, the air molecules lost energy as they pushed out the water and air. This means that the air in the bottle cooled off. (This is an adiabatic process—in which rapid expansion of the air caused it to cool off. This is the opposite case of the adiabatic process in the Fire Syringe in Chapter 3 in which the air became warmer because it was compressed.) If the air in the bottle cools off enough, it will reach the dew point, and fog will form in the bottle. As this happens, the relative humidity in the bottle increases, while the amount of vapor remains the same. When the relative humidity reaches 100% a cloud forms inside the bottle.

Web Resources

Learn about how pressure changes in air and water and predict how the pressure will change in various circumstances.
http://phet.colorado.edu/en/contributions/view/3569

Learn how the properties of gas (volume, heat, etc.) vary in relation to each other.
http://phet.colorado.edu/en/simulation/gas-properties

Exercises dealing with fluid pressure and depth.
www.grc.nasa.gov/www/K-12/WindTunnel/Activities/fluid_pressure.html

Activity on relationships among altitude, air density, temperature, and pressure.
www.grc.nasa.gov/www/K-12/problems/Jim_Rinella/AltitudevsDensity_act.htm

Activity on relationship between air pressure and temperature.
www.grc.nasa.gov/www/K-12/Missions/Rhonna/pre_act.htm

Control a piston in a chamber to explore relations among pressure, temperature, density, and volume.
http://jersey.uoregon.edu/vlab/Piston/

Simulation to investigate how pressure changes in air and water.
http://phet.colorado.edu/en/simulation/under-pressure

Questions to go with the PhET "Under Pressure" simulation.
http://phet.colorado.edu/en/contributions/view/3611

Fluid pressure activity.
http://phet.colorado.edu/en/contributions/view/3569

Relevant Standards

Note: The Next Generation Science Standards *can be viewed online at* www.nextgenscience.org/next-generation-science-standards.

PERFORMANCE EXPECTATIONS

MS-PS2-2

Plan an investigation to provide evidence that the change in an object's motion depends on the sum of the forces on the object and the mass of the object. [Clarification Statement: Emphasis is on balanced (Newton's First Law) and unbalanced forces in a system, qualitative comparisons of forces, mass and changes in motion (Newton's Second Law), frame of reference, and specification of units.] [Assessment Boundary: Assessment is limited to forces and changes in motion in one-dimension in an inertial reference frame, and to change in one variable at a time. Assessment does not include the use of trigonometry.]

SCIENCE AND ENGINEERING PRACTICES

Developing and Using Models

Modeling in 6–8 builds on K–5 and progresses to developing, using and revising models to describe, test, and predict more abstract phenomena and design systems.

- Develop a model to predict and/or describe phenomena. (MS-PS1-1),(MS-PS1-4)

- Develop a model to describe unobservable mechanisms. (MS-PS1-5)

Analyzing and Interpreting Data

Analyzing data in 6–8 builds on K–5 and progresses to extending quantitative analysis to investigations, distinguishing between correlation and causation, and basic statistical techniques of data and error analysis.

- Analyze and interpret data to determine similarities and differences in findings. (MS-PS1-2)

Obtaining, Evaluating, and Communicating Information

Obtaining, evaluating, and communicating information in 6–8 builds on K–5 and progresses to evaluating the merit and validity of ideas and methods.

- Gather, read, and synthesize information from multiple appropriate sources and assess the credibility, accuracy, and possible bias of each publication and methods used, and describe how they are supported or not supported by evidence. (MS-PS1-3)

CONNECTIONS TO NATURE OF SCIENCE

Science Models, Laws, Mechanisms, and Theories Explain Natural Phenomena

- Theories and laws provide explanations in science.

- Laws are statements or descriptions of the relationships among observable phenomena.

DISCIPLINARY CORE IDEAS

PS1.A: Structure and Properties of Matter

- Gases and liquids are made of molecules or inert atoms that are moving about relative to each other. (MS-PS1-4)

PS2.A: Forces and Motion

- For any pair of interacting objects, the force exerted by the first object on the second object is equal in strength to the force that the second object exerts on the first, but in the opposite direction (Newton's third law). (MS-PS2-1)

- The motion of an object is determined by the sum of the forces acting on it; if the total force on the object is not zero, its motion will change. The greater the mass of the object, the greater the force needed to achieve the same change in motion. For any given object, a larger force causes a larger change in motion. (MS-PS2-2)

CROSSCUTTING CONCEPTS

Patterns

- Different patterns may be observed at each of the scales at which a system is studied and can provide evidence for causality in explanations of phenomena.

Cause and Effect

- Empirical evidence is required to differentiate between cause and correlation and make claims about specific causes and effects.

- Systems can be designed to cause a desired effect.

Systems and System Models

- When investigating or describing a system, the boundaries and initial conditions of the system need to be defined.

5

DENSITY AND BUOYANCY

According to legend, "Eureka!" was first heard more than 2,000 years ago when the Greek mathematician Archimedes saw that the water level rose in the bathtub when he stepped into it. This made him realize that he could determine his body's volume if he knew the amount of water his body displaced in the tub. Archimedes went on to determine that a body immersed in fluid has a buoyant force equal to the weight of the fluid that it displaces. This fluid can be either a liquid or a gas.

The story goes that the king doubted whether his crown was made of pure gold. Archimedes was able to use his new method to find the truth: He said that the irregularly shaped crown would displace an amount of water equal to its volume. The equipment for measuring the mass already existed, so after measuring the volume, Archimedes was able to calculate the mass-to-volume ratio. This gave them the density of the metal in the crown, which could then be compared to the density of pure gold. Archimedes discovered that the goldsmith who had manufactured the crown had substituted some of the gold with silver—a cheaper metal. Thus, Archimedes used the concepts of density and buoyancy when finding out what the crown was made of.

You can use water to help familiarize yourself with the concepts of density and buoyancy. For example, lifting rocks from the bottom of a lake is easier than moving them on the beach. While in the water, the buoyant force makes the rocks feel lighter. Life jackets provide enough buoyancy to prevent a fisherman from sinking in the water and can also make diving impossible. A boat's anchor, made of steel, sinks, but the boat itself, also made of steel, has so much buoyant force acting on it that it can bear the weight of many passengers and of itself without sinking.

In this chapter, you will explore density and buoyancy.

KEYWORDS

Knowing these terms will help you to enjoy the explorations.

floating

sinking

mass

weight

volume

density

force

FIGURE 5.1: Two spheres

SAFETY NOTES

- Make sure the area is clear of anything fragile or breakable.

- Immediately wipe up any splashed water to prevent a slip or fall hazard.

A SINKING FEELING

In this experiment, you will explore how the density of an object, in this case a sphere (Figure 5.1), depends on its volume and mass.

Procedure

Take both spheres in your hands and explore them, using the questions to guide you.

Questions

- Based only on how the spheres feel in your hands, which sphere would you say is heavier?

- After you make a prediction, place the spheres on a scale and check their weights. Was your prediction correct?

- Will either of the spheres float in water? Make a prediction before trying.

- Calculate the density of the smaller sphere. *Hint*: You can use a measuring cup (or a graduated cylinder) and a scale.

- Based on the density, guess what material the sphere is made of.

ROCK THE BOAT

Why do boats float? They are supported by a force called buoyancy. With the Boat and Rock (Figure 5.2) you will experiment with different objects to learn more about buoyancy. You will need some extra equipment including a 1 kg weight or stone and a piece of wood.

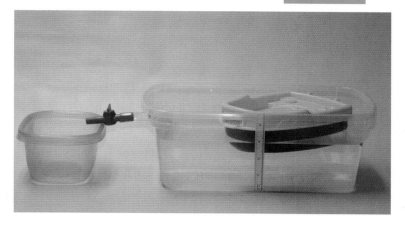

FIGURE 5.2: The Boat and Rock experiment.

Procedure

1. Put the boat in the container filled with water.

2. Test if the stone floats.

3. Put the stone or 1-kg weight into the boat so that the boat still floats and is in balance. If the 1-kg weight makes the boat sink, try a 1/2-kg weight.

4. Mark where the water level is on both the boat and on the side of the container. You can use a small piece of tape or a marker.

5. Predict what will happen to the water level—both compared with the boat and with the container—if the stone is moved from the boat into the water. Explain your prediction carefully before trying.

6. Repeat the experiment, replacing the stone with the piece of wood. Explain your prediction about the change in water levels before trying it.

Questions

*   How does the draft of the boat (how deep it is in the water) change if different weights are used? Why?

*   Explain any changes in water level in the container when the stone or the wood are moved into the water.

Additional Experiment

*   Starting with an empty overflow container, put the stone on the boat and compare the weight of the overflow water with the weight of the stone.

▼
SAFETY NOTES

*   Wear safety glasses or goggles.

*   Immediately wipe up any splashed water to prevent a slip or fall hazard.

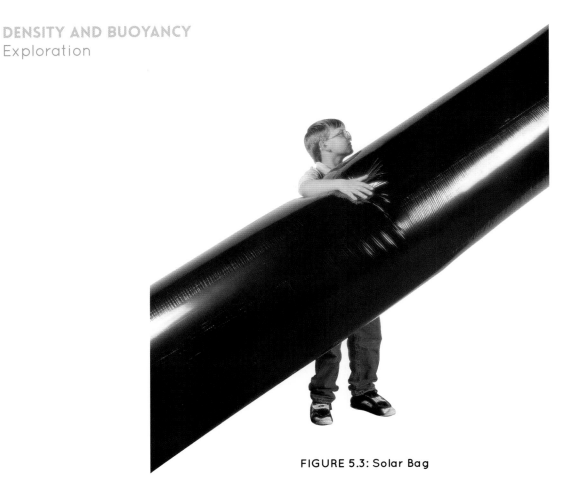

FIGURE 5.3: Solar Bag

SOLAR BAG

Buoyancy affects objects in all fluids—not only in water or other liquids, but also in gases. The Solar Bag (Figure 5.3) demonstrates the effect of buoyancy in the air.

Procedure

1. Take the Solar Bag out on a sunny day.

2. Inflate the Solar Bag.

3. Tie one end of the bag to a cord and make sure the bag cannot get loose and be blown away.

4. Put the bag in a sunny spot and watch for something to happen.

Questions

• Why do you think it is called a Solar Bag?

• Describe what happens after the bag has been in direct sunlight for a while. Why does it happen?

• It is said that "warm air is lighter than cold air." Explain this from a physics point of view.

• Where is this type of phenomenon used, or where could it be used?

CARTESIAN DIVER

In this activity, you will examine buoyancy and how different substances respond to changes in pressure using the Cartesian Diver (Figure 5.4).

Procedure

1. Examine the diver, and then put a small amount of water into it.

2. Drop the diver into a plastic bottle filled with water. The amount of water in the diver should allow it to just barely float.

3. Squeeze the bottle, somewhat hard if necessary, to get the diver to "swim" to the bottom.

4. When the diver has swum to the bottom of the bottle, you can release the bottle and let the diver swim up.

5. Observe any changes in the diver when the bottle is squeezed.

Questions

*   Explain how the Cartesian Diver works. Why does it dive?

*   Why and how does the density of the diver change when you compress the bottle?

FIGURE 5.4: Cartesian Diver

▼
SAFETY NOTES

*   Wear safety glasses or goggles.

*   Immediately wipe up any splashed water to prevent a slip or fall hazard.

A SINKING FEELING

With the two spheres (Figure 5.5) from the Steel Sphere Density Kit, you were able to explore the relations between mass and volume in terms of density. Density is a characteristic that describes how tightly packed the matter is in an object. Density can also be described as the mass per unit volume. Mathematically, density can be expressed as

$$\rho = \frac{m}{V}$$

where ρ is density, m is mass, and V is volume. The units of density in SI units are kg/m^3. Sometimes the units g/cm^3 are used instead.

FIGURE 5.5: Two steel spheres

The smaller sphere may seem heavier in your hand because the pressure on your hand is greater. According to Archimedes' principle, the buoyant force on an object is equal to the weight of the fluid displaced by the object. This upward force is called *buoyancy*. An object floats if the buoyant force is equal to the weight of the object.

In this experiment, the smaller sphere sinks because the weight of the water displaced by the sphere is not enough to provide sufficient buoyant force. The buoyancy is less than the gravitational force. We can also compare the density of water to the density of the sphere. The sphere sinks if the density of the sphere is greater than that of the water. The bigger sphere floats because its density is less than that of the water.

The formula for the volume of a sphere involves the radius of the sphere. However, it is easier to measure the diameter or circumference of a sphere than the radius. Therefore, you can measure one of those other quantities first, and then calculate the radius. After you have found the radius, r, you can calculate the volume, V, of the sphere using this formula:

$$V = \frac{4\pi r^3}{3}$$

Now, if you have the mass and volume of the sphere, you can easily calculate the ratio of the two, which is the sphere's density. Once you have determined the density, you can guess that the spheres are made of steel, which has an approximate density of 7,750 kg/m^3.

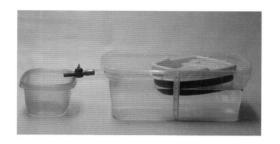

FIGURE 5.6: Boat and Rock

ROCK THE BOAT

When you place the boat of the Boat and Rock (Figure 5.6) set in the water for the "Rock the Boat" exploration, it floats. Conversely, the stone or weight sinks. When you place the stone into the boat, the draft of the boat is bigger—the boat is now loaded and is displacing more water than before. According to Archimedes' principle, the weight of the water displaced is equal to the combined weight of the boat and the stone. The water level in the container is now higher than in the beginning (before you put the stone in the boat).

When the stone is moved from the boat to the container, the draft of the boat decreases, that is, the boat isn't as far down in the water as it was before. If you then move the stone to the bottom of the container, it displaces an amount of water with the same volume as itself, which is less than the amount of water that was displaced when the stone was in the boat. Remember, when the stone is in the boat, the additional water displaced has the same weight as the stone. Since water is less dense than stone, the displaced water must have a volume greater than that of the stone, and the water level goes down. If you do the experiment with a piece of wood, you will see that the water level remains pretty much the same because the wood is essentially floating whether it is in the boat or in the water.

Additional experiment: Starting with an empty overflow container, put the stone on the boat and compare the weight of the overflow water with the weight of the stone.

SOLAR BAG

When you take the Solar Bag (Figure 5.7) outside, the Sun heats up the air inside it. Heating the air makes the molecules move faster and collide with the inner walls of the bag more forcefully and frequently. As a result, the pressure rises, making the air (and bag) expand. Although the amount of air in the bag remains the same, its volume increases. Eventually, the density of air inside the bag decreases and becomes less than the density of the air outside the bag. This is when buoyancy gets involved.

The bag's buoyancy is equal to the weight of the medium displaced, so when the air expands inside the bag, the bag displaces a bigger volume of the outside cooler and heavier air. The weight of the displaced cooler air is greater than that of the combined weight of the bag and the air inside of it. In other words, the buoyancy is greater than the gravitational force pulling down on the bag. This is why the bag rises in the air.

FIGURE 5.7: Solar Bag

FIGURE 5.8: Cartesian Diver

CARTESIAN DIVER

When you squeeze the bottle with the Cartesian Diver (Figure 5.8) inside, the pressure rises in the bottle. Water is an incompressible fluid, but the air inside the diver itself can be compressed. As the pressure compresses the air in the diver, water flows into the diver, increasing its weight.

In accordance with Archimedes' principle, when the weight of the diver is greater than its buoyancy, the diver sinks to the bottom of the bottle. When you stop squeezing, the air expands back to its initial condition, water flows out of the diver, the density of the diver decreases, and the diver rises.

In order to make the diver hover in the water you must squeeze the bottle with just the right, constant force. In this way, you can make the buoyancy and the weight equal. In a force diagram, you would draw force vectors that are oppositely directed but have the same length.

Buoyancy can be explained by hydrostatic pressure. Hydrostatic pressure is the pressure caused by the weight of the fluid itself. For example, the hydrostatic pressure a diver feels in 1-meter-deep water is approximately 10,000 pascals (Pa). This is the pressure above the normal atmospheric pressure at the surface of the water, which is approximately 100,000 Pa. At the bottom of a 10-meter-deep pool, the hydrostatic pressure is approximately 100,000 Pa—about the same as atmospheric pressure. So the total pressure at this depth is 200,000 Pa—of which one half (100,000 Pa) is from atmospheric pressure and the other half (100,000 Pa) is from the weight of the water above the diver.

The deeper in the liquid an object is, the greater the pressure. This means that the pressure on the bottom of an object is greater than the pressure on the top. The pressure difference between the top and bottom causes the buoyancy. Buoyancy is the force that holds up a boat, lifts a balloon in the air, and makes a bubble rise in a soda bottle.

The buoyant force, N, can be calculated by

$$N = \rho V g$$

where ρ is the density of the fluid, V is the volume of displaced fluid, and g is the acceleration of gravity. In short, buoyancy is equal to the weight of the fluid displaced by the object. The buoyancy also depends on the density of the fluid.

Web Resources

Learn about buoyancy by floating blocks in liquid.
http://phet.colorado.edu/en/simulation/buoyancy

Explore how size, weight, and density affect an object's buoyancy.
http://phet.colorado.edu/en/simulation/density

See how much water can be added to a vessel before it sinks. Monitor the mass, density, and volume as you go.
www.mhhe.com/physsci/physical/giambattista/buoyancy/buoyancy.html

Play with the effects of buoyancy in liquid.
www.walter-fendt.de/ph14e/buoyforce.htm

Experiment with the Cartesian Diver to learn about Archimedes' principle as well as the ideal gas law.
http://lectureonline.cl.msu.edu/~mmp/applist/f/f.htm

Experiment with Galileo's thermometer: estimate the temperature by looking at the colored beads.
http://scratch.mit.edu/projects/27574/

Relevant Standards

Note: The Next Generation Science Standards *can be viewed online at* www.nextgenscience.org/next-generation-science-standards.

PERFORMANCE EXPECTATIONS

MS-PS1-4

Develop a model that predicts and describes changes in particle motion, temperature, and state of a pure substance when thermal energy is added or removed. [Clarification Statement: Emphasis is on qualitative molecular-level models of solids, liquids, and gases to show that adding or removing thermal energy increases or decreases kinetic energy of the particles until a change of state occurs. Examples of models could include drawings and diagrams. Examples of particles could include molecules or inert atoms. Examples of pure substances could include water, carbon dioxide, and helium.]

SCIENCE AND ENGINEERING PRACTICES

Developing and Using Models

Modeling in 6–8 builds on K–5 and progresses to developing, using and revising models to describe, test, and predict more abstract phenomena and design systems.

- Develop a model to predict and/or describe phenomena. (MS-PS1-1),(MS-PS1-4)

- Develop a model to describe unobservable mechanisms. (MS-PS1-5)

Analyzing and Interpreting Data

Analyzing data in 6–8 builds on K–5 and progresses to extending quantitative analysis to investigations, distinguishing between correlation and causation, and basic statistical techniques of data and error analysis.

- Analyze and interpret data to determine similarities and differences in findings. (MS-PS1-2)

Obtaining, Evaluating, and Communicating Information

Obtaining, evaluating, and communicating information in 6–8 builds on K–5 and progresses to evaluating the merit and validity of ideas and methods.

- Gather, read, and synthesize information from multiple appropriate sources and assess the credibility, accuracy, and possible bias of each publication and methods used, and describe how they are supported or not supported by evidence. (MS-PS1-3)

CONNECTIONS TO NATURE OF SCIENCE

Science Models, Laws, Mechanisms, and Theories Explain Natural Phenomena

- Theories and laws provide explanations in science.

- Laws are statements or descriptions of the relationships among observable phenomena.

DISCIPLINARY CORE IDEAS

PS3.A: Definitions of Energy

- The term "heat" as used in everyday language refers both to thermal energy (the motion of atoms or molecules within a substance) and the transfer of that thermal energy from one object to another. In science, heat is used only for this second meaning; it refers to the energy transferred due to the temperature difference between two objects. (secondary to MS-PS1-4)

- The temperature of a system is proportional to the average internal kinetic energy and potential energy per atom or molecule (whichever is the appropriate building block for the system's material). The details of that relationship depend on the type of atom or molecule and the interactions among the atoms in the material. Temperature is not a direct measure of a system's total thermal energy. The total thermal energy (sometimes called the total internal energy) of a system depends jointly on the temperature, the total number of atoms in the system, and the state of the material. (secondary to MS-PS1-4)

CROSSCUTTING CONCEPTS

Patterns

- Different patterns may be observed at each of the scales at which a system is studied and can provide evidence for causality in explanations of phenomena.

Cause and Effect

- Empirical evidence is required to differentiate between cause and correlation and make claims about specific causes and effects.

- Systems can be designed to cause a desired effect.

6

FORCE, MOTION, AND ENERGY

In order to discuss mechanical interactions, we need to understand forces and motion. In physical science, we can make use of all three of Newton's laws of motion in a variety of situations. You experience the first law when you're riding in a car or bus and the driver suddenly brakes or makes a sharp turn. The second law tells us why racing cars are made light, and why it's difficult to stop a big truck. Newton's third law tells us that forces always come in pairs.

Newton's third law about forces coming in pairs says that any interaction between two bodies causes equal but opposite forces on the bodies. The forces arise from the same interaction, but they act on different bodies. This can be easily demonstrated with two spring scales. When the scales are attached to each other and pulled apart, the readings in both scales are the same.

Energy analysis encompasses all areas of physics. In mechanics, energy analysis is all about potential and kinetic energy and their changes. According to the law of conservation of energy, the amount of energy in a closed system always remains the same—no energy is created and no energy is lost—the energy only transforms from one form to another. In some of the examples in this chapter, the gravitational potential energy (energy that arises from the height to which an object has been lifted) is transformed to kinetic energy (energy from the object's motion), and some of the energy is converted to another form by frictional forces.

KEYWORDS AND FORMULAS

You may want to refer to the following concepts and formulas for the explorations in this chapter:

force: $F = ma$

acceleration: $a = \dfrac{\Delta v}{\Delta t}$

force on an object due to gravity: (weight) $F_g = mg$

velocity: $v = \dfrac{d}{t}$

work: $W = Fd$

law of conservation of energy: $E(\text{initial}) = E(\text{final})$

kinetic energy: $E_k = \dfrac{1}{2}mv^2$

gravitational potential energy: $E_p = mgh$

Exploration

▼
SAFETY NOTE
Wear safety glasses or goggles.

EXCELLING AT ACCELERATION

With the help of the Pull-Back Car (Figure 6.1), you will learn more about a quantity called *acceleration*.

Procedure

1. Experiment to see how long of a track is needed for the car to reach its maximum velocity.

2. Plan a set of measurements to determine the acceleration of the car. There are many different ways to do this: a motion sensor, a tape timer, or just a ruler and a stopwatch. The kit also inlcudes a photogate timer that can be used.

3. Repeat the measurements with passengers (small weights) in the car.

Questions

- What is the acceleration of the car?

- How does the acceleration change if there are passengers in the car? Explain.

FIGURE 6.1: Pull-Back Car

BALANCING BIRD

In the previous exploration, you learned that if there is a net force, there will be an acceleration. If the net force is zero, an object will move at a constant velocity or remain at rest. This Balancing Bird (Figure 6.2) will show you an interesting way for an object to remain stationary.

SAFETY NOTE

Wear safety glasses or goggles.

Procedure

1. Balance the bird on something, such as a fingertip.
2. Try balancing from the wings, body, tail, beak, and so on.

Questions

- When is the bird in balance?
- Why are the wings spread?
- Explore the balance points of different objects. See in what ways they will be stable or unstable. What does it take for the object to have good balance and be stable?

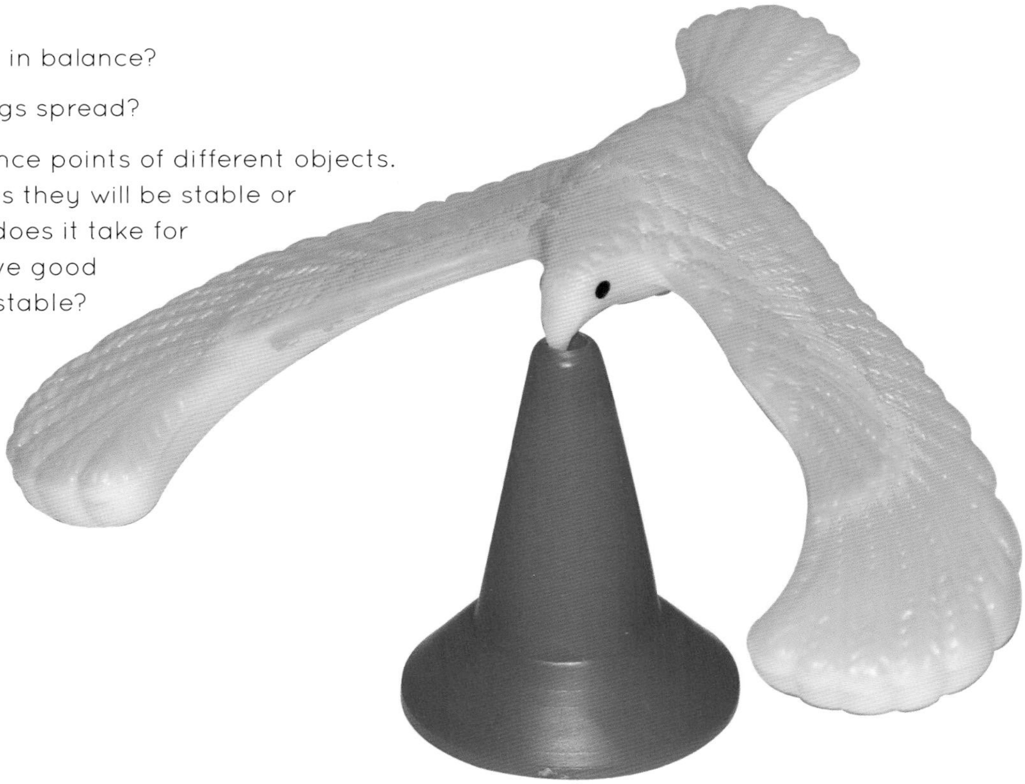

FIGURE 6.2: Balancing Bird

FORCE, MOTION, AND ENERGY
Exploration

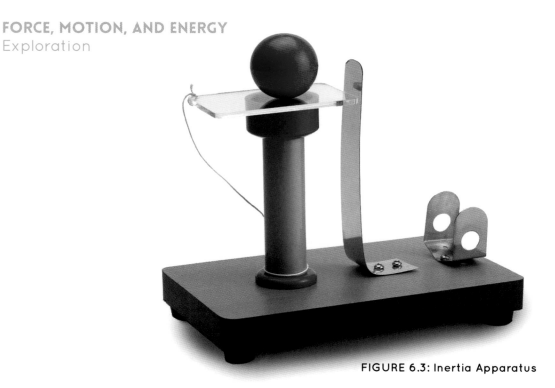

FIGURE 6.3: Inertia Apparatus

▼
SAFETY NOTE

Wear safety glasses or goggles.

IT'S A SNAP!

The Inertia Apparatus (Figure 6.3) can be used to explore Newton's first law, Newton's second law, and the concept of inertia.

Procedure

1. Put the plastic plate on the blue post.

2. Put the ball on the plate.

3. Pull back on the spring and then release it.

Questions

*   Predict what will happen when the metal strip hits the plastic plate.

*   After you do the experiment, record what happened.

*   Explain the physics involved with this phenomenon.

*   If you replace the ball with an eraser or a small weight, are the results different? If so, how?

RACE TO THE BOTTOM

With the Vertical Acceleration Demonstrator (VAD; Figure 6.4), you can explore the acceleration of falling objects.

Procedure

1. Put the VAD on a stand so that it is approximately 1.5 m above the floor.

2. Check that the VAD is level.

3. Put the balls in place, answer the first question, and then activate the VAD.

Questions

- Which ball will hit the ground first after you pull the trigger? Make a prediction before trying.

- Why do the balls fall in the first place?

- Which ball hit the ground first? Explain your observations.

▼ SAFETY NOTES

- Wear safety glasses or goggles.

- Use caution when balls are dropped on the floor as they can cause a slip, trip, or fall hazard.

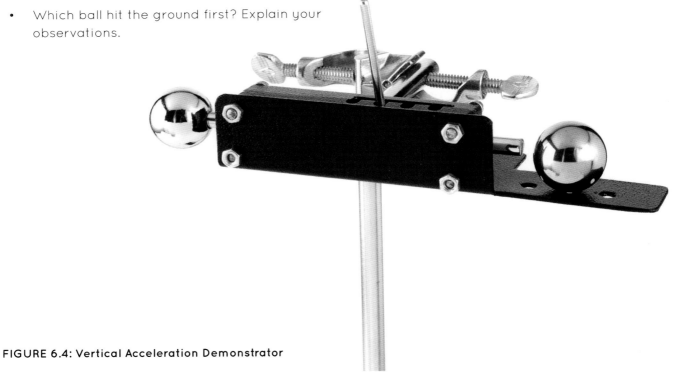

FIGURE 6.4: Vertical Acceleration Demonstrator

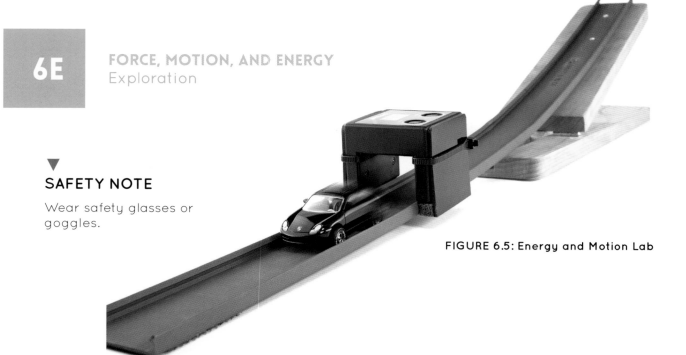

FIGURE 6.5: Energy and Motion Lab

▼
SAFETY NOTE

Wear safety glasses or
goggles.

ENERGY INGENUITY

**With the car track and photogate timer
from the Energy and Motion Lab (Figure
6.5), you can study the conservation
of energy. Your challenge here is to
determine the connection between
the height of the starting point and
the velocity at the end of the track.**

Procedure

1. Determine the mass of the car. (You
    might need this to answer one of the
    questions.)

2. Mark a starting line at the beginning
    of the track. This is how you make sure
    the distance remains the same in every
    run. Then, lift the beginning of the
    track approximately 10 cm (4 in.) for
    the first run.

3. Check that the photogate timer is on.
    Make sure that the timer is measuring
    speed—if not, press the button for
    two seconds. When the unit "m/s" is
    flashing on the screen, the photogate is
    ready to measure the speed.

4. Release the car at the starting line, and
    check the reading in the photogate
    timer.

5. Make notes of the results you get from
    different heights. (Use h = height, v =
    velocity.)

Questions

*   How does the velocity (v) at the end
    seem to depend on the initial height
    (h)?

*   Square the velocities (v^2), and make
    a graph of v^2 versus h, where v^2
    represents the y axis and h is on the
    x axis. What does the graph tell you
    about what is going on?

*   What is the form of the car's energy at
    the beginning?

*   Some of the energy is transformed to
    kinetic energy and some of it is lost
    because of friction. How much energy
    is lost because of frictional forces?

HAPPY / UNHAPPY BALLS

With the Happy / Unhappy Balls (Figure 6.6) you will explore various aspects of collisions and conservation of energy.

Procedure

1. Drop both balls on the floor at the same time. Record your observations.

2. Examine the balls closely. Is there any difference?

3. Roll the balls on an inclined plane.

Questions

- What did you notice when you dropped the balls on the floor? Why do you think this happened?

- Where did the kinetic energy in each ball go?

- Why is there a difference when rolling the balls?

▼
SAFETY NOTES

- Wear safety glasses or goggles.

- Use caution when balls are dropped on the floor as they can cause a slip, trip, or fall hazard.

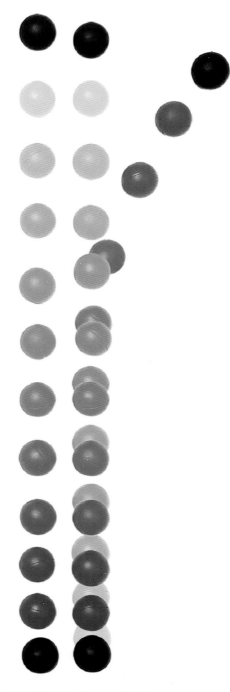

FIGURE 6.6: Happy / Unhappy Balls

FIGURE 6.7: Pull-Back Car

EXCELLING AT ACCELERATION

The word *acceleration* is often used only when speed is increasing, but actually an object is accelerating every time its speed or direction changes. When the speed is decreasing, the acceleration is negative. The Pull-Back Car (Figure 6.7) allows you to explore factors that affect acceleration.

One thing that affects the acceleration is the mass. You probably found out that adding mass or a passenger to the car decreased acceleration. The more mass an object has, the more difficult it is for it to speed up or slow down. When building a race car, the builders tend to make the car as light as possible so it will have greater acceleration. A high mass is the reason heavy trucks need a longer distance to brake.

All this is about Newton's second law,

$$F = ma$$

where F is the force, m is the mass and a is the acceleration of the object.

BALANCING BIRD

When an object such as the Balancing Bird (Figure 6.8) is supported from its balance point, it stays in balance. You can imagine that the balance point is a point where all the mass of the object is concentrated.

If you redistribute the mass in an object, the balance point might change to a different place. This is why it's advisable to load all the heavy stuff on the bottom and the lighter stuff up higher in a large truck or moving van. That way, you get the balance point lower and the van will not tip over as easily.

Another thing that substantially contributes to the balance of an object is the amount of supporting area. For example, the larger the area between four chair legs is, the better the chair stays in balance and does not tip over. One condition for balance is that an object's balance point must be somewhere over the supporting area. If the balancing point is not over the supporting area, the object is not in balance and will fall over. The balance point can also be demonstrated with an empty shoe box and weights or stones. Put the box on the edge of the table and place the weights so that the box remains in balance even though more than half the box is beyond the edge of the table.

The Balancing Bird's balance point is on its beak. This is because weight is added at the end of the bird's wings, which are spread wide and extend forward.

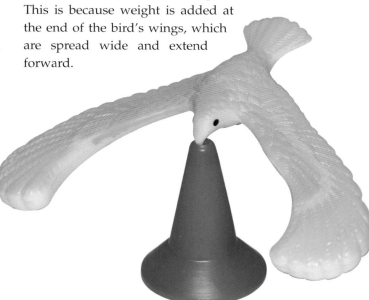

FIGURE 6.8: Balancing Bird

IT'S A SNAP!

With the Inertia Apparatus (Figure 6.9) you can see a phenomenon you experience in everyday life: inertia. For example, when traveling in an accelerating car, passengers feel pushed back in their seats; and when the car is braking, passengers tend to lean forward. When the car takes a turn, passengers feel themselves being pushed toward the outside of the curve. The passengers keep moving at the same speed and in the same direction as they had been before the change.

The greater an object's mass, the harder it is to change the object's speed and direction. When a massive object is moving, it is difficult to stop. If a massive object is stationary, it is difficult to get it moving very fast. To make an object with mass, m, move with acceleration, a, requires a force, F, given by

$$F = ma.$$

This is Newton's second law of motion. The object's mass is a measure of its inertia, that is, how strongly it resists changes in motion.

In the Inertia Apparatus, the ball stays almost still even though the plastic plate under it is hit with the metal strip. The forces affecting the ball are not strong enough to make it move much. The plate, however, takes off quickly because it is much lighter than the ball and receives a stronger force than the ball does. The plate has less mass than the ball and therefore less inertia.

Newton's first law explains what happens when there is no net force on an object. Thus, it applies when the ball is just sitting on the plate before the plate is snapped out. Once the plate is moving, there is a (small) frictional force on the ball. Now that there is a nonzero net force, you should think about Newton's second law, not his first law. Because the frictional force acting on the

FIGURE 6.9: Inertia Apparatus

ball is small and acts for a very short time (thanks to the plate's low mass and Newton's second law indicating that it will have a high acceleration), the horizontal acceleration of the ball is very low. So it doesn't move far enough horizontally to fall off the post. Once the plate is gone and the ball is stationary on the post, the net force is again zero, and Newton's first law applies again.

This phenomenon can also be demonstrated with a thin thread tied to a weight. If the thread is pulled quickly, it breaks. If the thread is pulled slowly, it does not break, and it is possible to make the weight move. Another example that involves the same phenomenon as the Inertia Apparatus is the tablecloth trick, in which the magician pulls a tablecloth out from under a vase or some dishes. For this to work, the magician must have quick hands and the dishes must be heavy enough. (Check the Web Resources for a link to see the tablecloth trick performed by one of the authors.)

RACE TO THE BOTTOM

The Vertical Acceleration Demonstrator (Figure 6.10) allows you to play more with Newton's second law. According to this law, when an object with mass, m, experiences forces on it with the sum ΣF, it accelerates with an acceleration, a, given by

$$a = \frac{\Sigma F}{m}$$

The total force and the acceleration are directly proportional quantities, so the greater the force, the greater the acceleration. The downward acceleration of the balls due to the Earth's gravity is approximately 9.8 m/s². The acceleration of the two balls is the same because they have the same mass and experience the same force from the Earth's gravity. Therefore, they hit the ground at the same time. The horizontal movement does not affect the time to fall, because the vertical and horizontal components of the balls' motion are completely independent.

The gravitational force is equal to the weight, F_g, of the object,

$$F_g = mg,$$

where m is the mass, and g is the acceleration due to gravity.

FIGURE 6.10: Vertical Acceleration Demonstrator

ENERGY INGENUITY

In the Energy and Motion Lab (Figure 6.11), you explore the concept that gravitational potential energy arises from the interaction between an object and the Earth. The potential energy that an object has depends on the object's weight and the height of the body. You can calculate the potential energy using

$$E = mgh$$

where m is the mass, g is the gravitational acceleration, and h is the height from some chosen zero level, such as ground level or table height.

When an object is on the move, it has kinetic energy. Its kinetic energy depends on the object's mass and its velocity. The more massive the object and the faster it moves, the more kinetic energy it has.

According to the law of conservation of energy, the amount of energy in a closed system always remains the same. However, energy can be transformed from one form to another. In these experiments, most of the potential energy is transformed to kinetic energy. When the car is still at the starting line, it has only potential energy. When the car has been released, it accelerates, and most of the potential energy transforms to kinetic energy. A small part of the energy also gets transformed to thermal

energy due to friction, air resistance, and sound. If these are small, then the square of the velocity, v^2, is almost directly proportional to the initial height, h.

If there weren't any frictional forces or other losses of energy, all of the potential energy would be transformed to kinetic energy.

$$mgh = \frac{1}{2}mv^2.$$

Here, a small part of the potential energy is transformed to work done by friction and air resistance. Ignoring the insignificant loss due to sound, the work (W) done by these two forces is the difference between the initial potential energy and the final kinetic energy.

$$W = mgh - \frac{1}{2}mv^2$$

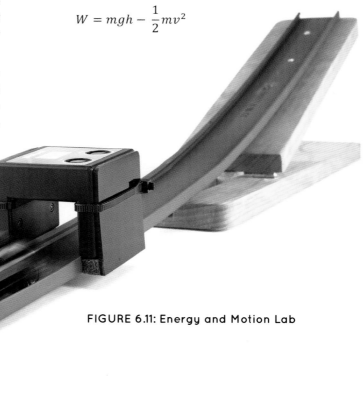

FIGURE 6.11: Energy and Motion Lab

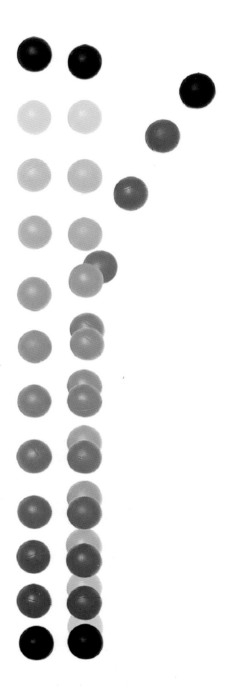

FIGURE 6.12: Happy / Unhappy Balls

HAPPY / UNHAPPY BALLS

The two balls in the Happy / Unhappy Balls set (Figure 6.12) seem to be similar. However, they are made of different materials. You can see the difference when trying to bounce or roll the balls. One ball, the "happy" one, bounces high and rolls nicely, while the other, the "unhappy" ball, does not bounce or roll well.

Perhaps you are not surprised that a rubber ball bounces well. Some kinds of rubber can be very elastic. The two balls respond differently when they hit the floor because they are made of different blends of rubber. At first, before the balls are released, both balls have the same amount of potential energy. Some of that energy gets converted to kinetic energy and some gets converted to thermal energy by internal friction—friction between the chains of molecules in the rubber blend.

The blend used in the happy ball makes it possible for most of the potential energy stored in the elastic molecules to be converted to kinetic energy. The unhappy ball doesn't store energy as potential energy in elastic molecules. Rather, energy in the unhappy ball gets converted to thermal energy as a result of friction between its molecules. Thus, the happy ball's bounces are elastic and the unhappy ball's bounces are almost completely inelastic. Further, when the balls roll down a small incline, the happy ball rolls better than the unhappy one. When balls roll along a surface, they get slightly deformed. (Think of how the bottom of a car tire gets flattened where it touches the pavement.) When the happy ball rolls, the flattened part "bounces back" with little energy loss. But when the unhappy ball rolls, the energy used to flatten it at the bottom gets lost as heat, so that ball loses energy faster.

Web Resources

Basic concepts and terminology about kinetic energy.
www.physicsclassroom.com/Class/energy/U5L1c.cfm

Learn about kinetic energy, potential energy, and friction by creating simulations of skateboarding ramps.
http://phet.colorado.edu/en/simulation/energy-skate-park-basics

Take your skateboarding simulations to another level to learn more about conservation of energy. Change things up by going to different planets or outer space.
http://phet.colorado.edu/en/simulation/energy-skate-park

See how the kinetic and potential energy change as a pendulum swings back and forth.
www.mta.ca/faculty/science/physics/ntnujava/Pendulum/Pendulum.html

Explore elastic collisions of balls in two dimensions. Change the mass ratio and the impact parameter.
http://physics.usask.ca/~pywell/p121/Notes/collision/collision.html

The tablecloth trick demonstrated by Dr. Matt Bobrowsky.
www.msb-science.com/tablecloth-trick.mov

Relevant Standards

Note: The Next Generation Science Standards *can be viewed online at* www.nextgenscience.org/next-generation-science-standards.

PERFORMANCE EXPECTATIONS

MS-PS2-2

Plan an investigation to provide evidence that the change in an object's motion depends on the sum of the forces on the object and the mass of the object. [Clarification Statement: Emphasis is on balanced (Newton's first law) and unbalanced forces in a system, qualitative comparisons of forces, mass and changes in motion (Newton's second law), frame of reference, and specification of units.] [Assessment Boundary: Assessment is limited to forces and changes in motion in one-dimension in an inertial reference frame, and to change in one variable at a time. Assessment does not include the use of trigonometry.]

MS-PS2-4

Construct and present arguments using evidence to support the claim that gravitational interactions are attractive and depend on the masses of interacting objects. [Clarification Statement: Examples of evidence for arguments could include data generated from simulations or digital tools; and charts displaying mass, strength of interaction, distance from the Sun, and orbital periods of objects within the solar system.] [Assessment Boundary: Assessment does not include Newton's Law of Gravitation or Kepler's Laws.]

MS-PS3-1

Construct and interpret graphical displays of data to describe the relationships of kinetic energy to the mass of an object and to the speed of an object. [Clarification Statement: Emphasis is on descriptive relationships between kinetic energy and mass separately from kinetic energy and speed. Examples could include riding a bicycle at different speeds, rolling different sizes of rocks downhill, and getting hit by a wiffle ball versus a tennis ball.]

MS-PS3-2

Develop a model to describe that when the arrangement of objects interacting at a distance changes, different amounts of potential energy are stored in the system. [Clarification Statement: Emphasis is on relative amounts of potential energy, not on calculations of potential energy. Examples of objects within systems interacting at varying distances could include: the Earth and either a roller coaster cart at varying positions on a hill or objects at varying heights on shelves, changing the direction/orientation of a magnet, and a balloon with static electrical charge being brought closer to a classmate's hair. Examples of models could include representations, diagrams, pictures, and written descriptions of systems.] [Assessment Boundary: Assessment is limited to two objects and electric, magnetic, and gravitational interactions.]

MS-PS3-5

Construct, use, and present arguments to support the claim that when the kinetic energy of an object changes, energy is transferred to or from the object. [Clarification Statement: Examples of empirical evidence used in arguments could include an inventory or other representation of the energy before and after the transfer in the form of temperature changes or motion of object.] [Assessment Boundary: Assessment does not include calculations of energy.]

SCIENCE AND ENGINEERING PRACTICES

Analyzing and Interpreting Data

Analyzing data in 6–8 builds on K–5 and progresses to extending quantitative analysis to investigations, distinguishing between correlation and causation, and basic statistical techniques of data and error analysis.

- Construct and interpret graphical displays of data to identify linear and nonlinear relationships. (MS-PS3-1)

Constructing Explanations and Designing Solutions

Constructing explanations and designing solutions in 6–8 builds on K–5 experiences and progresses to include constructing explanations and designing solutions supported by multiple sources of evidence consistent with scientific ideas, principles, and theories.

- Apply scientific ideas or principles to design, construct, and test a design of an object, tool, process or system. (MS-PS3-3)

Engaging in Argument From Evidence

Engaging in argument from evidence in 6–8 builds on K–5 experiences and progresses to constructing a convincing argument that supports or refutes claims for either explanations or solutions about the natural and designed worlds.

- Construct, use, and present oral and written arguments supported by empirical evidence and scientific reasoning to support or refute an explanation or a model for a phenomenon. (MS-PS3-5)

CONNECTIONS TO NATURE OF SCIENCE

Science Models, Laws, Mechanisms, and Theories Explain Natural Phenomena

- Theories and laws provide explanations in science.

- Laws are statements or descriptions of the relationships among observable phenomena.

DISCIPLINARY CORE IDEAS

PS2.A: Forces and Motion

- For any pair of interacting objects, the force exerted by the first object on the second object is equal in strength to the force that the second object exerts on the first, but in the opposite direction (Newton's third law). (MS-PS2-1)

- The motion of an object is determined by the sum of the forces acting on it; if the total force on the object is not zero, its motion will change. The greater the mass of the object, the greater the force needed to achieve the same change in motion. For any given object, a larger force causes a larger change in motion. (MS-PS2-2)

PS3.A: Definitions of Energy

- Motion energy is properly called kinetic energy; it is proportional to the mass of the moving object and grows with the square of its speed. (MS-PS3-1)

- A system of objects may also contain stored (potential) energy, depending on their relative positions. (MS-PS3-2)

PS3.B: Conservation of Energy and Energy Transfer

- When the motion energy of an object changes, there is inevitably some other change in energy at the same time. (MS-PS3-5)

CROSSCUTTING CONCEPTS

Energy and Matter

- Changes of energy and matter in a system can be described in terms of energy and matter flows into, out of, and within that system.

- Energy cannot be created or destroyed—only moves between one place and another place, between objects and/or fields, or between systems.

Energy and Matter

- Energy may take different forms (e.g., energy in fields, thermal energy, energy of motion). (MS-PS3-5)

- The transfer of energy can be tracked as energy flows through a designed or natural system. (MS-PS3-3)

7

IT'S SCIENCE, NOT MAGIC

In this chapter, we will show you gadgets and gizmos that make science look like magic. You don't have to be a magician to perform these tricks—all you need is physics. Each student can pick a gizmo from this chapter and prepare a performance for the rest of the group. Remember that a real magician never tells the secrets of the tricks, so make your group think—don't give answers right away. It's definitely helpful for the magical physicist to study the trick and the physics before standing up on stage.

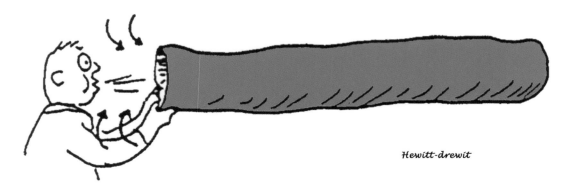

Hewitt-drewit

FIGURE 7.1: "Bernoulli's Bag"

WIND BAG

With the wind bag, often sold as "Bernoulli's Bag" (Figure 7.1), you will learn about an interesting phenomenon that allows you to inflate a bag much faster than you might otherwise expect.

Procedure

1.   Take two to four Bernoulli's Bags and tie a knot in one end of each tube.

2.   Place the bags next to each other on the floor.

3.   Have a competition to inflate the bags—the student who does it fastest wins. One student can be the referee.

4.   Next, the referee should inflate the bag with only a couple of breaths (see the Analysis for directions).

Questions

- What did you do with the bag when you performed the experiment? What happened?

- What happened when the referee filled the bag? Why?

- Where else might this phenomenon be used?

MIRAGE

The name Mirage (Figure 7.2) describes the gadget quite well. With Mirage you can study images created in mirrors and discover the difference between a virtual and a real image.

FIGURE 7.2: Mirage

Procedure

1. Put a small object in the middle of the lower mirror.

2. Put the top on.

3. Choose a spot where you can see the image on the top. Viewing from some angles works better than from other angles.

4. Ask the viewers to grab the object on the Mirage.

Questions

Web searches will probably help to answer these questions:

- Explain the concepts (a) *virtual image* and (b) *real image*.

- Is the image in the Mirage virtual or real? Why?

- What might be a practical use of this phenomenon?

▼
SAFETY NOTES

- Wear safety glasses or goggles.

- Use caution when handling the Mirage. Glass components can shatter if dropped and cut skin.

FIGURE 7.3: Dropper Popper

▼
SAFETY NOTE

Wear safety glasses or
goggles.

DROPPER POPPER

According to the law of conservation of energy, the
amount of energy in an isolated (or closed) system
always remains the same. Energy does not increase, and
it never disappears. However, this law may seem false
where the Dropper Poppers (Figure 7.3) are concerned.

Procedure

1. Prepare the Popper by turning its edges down.

2. Drop the Popper from a height of 1 meter. You can also
    try other heights. Be careful: The bounce is sometimes
    unpredictable.

Questions

*   What did you notice?

*   It seems that the Popper suddenly gets more energy.
    Why is that? Where does the extra energy come from?

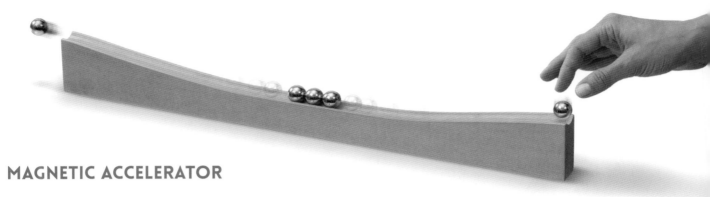

MAGNETIC ACCELERATOR

In nature, the amount of energy in an isolated (or closed) system does not increase or decrease. The energy can only be converted from one form of energy to another. In this activity with the Magnetic Accelerator (Figure 7.4), like the last one, it may seem that the law of conservation of energy is not applicable.

FIGURE 7.4: Magnetic Accelerator

Procedure

1. Set aside the magnetic ball.

2. Put one of the other balls at the top of the track and let it go.

3. When the first ball has settled on the bottom of the track, put the second ball at the top of the track and let it go.

4. Repeat this with the third ball.

5. Use the magnetic ball as the fourth ball.

6. Finally, place the fifth ball at the top of the track and let it go.

Questions

*   What did you observe?

*   What happened to the potential energy from the first three balls?

*   How do you explain what happened with the last two balls?

*   Did the ball get more energy somehow? Explain your ideas.

▼
SAFETY NOTES

*   Wear safety glasses or goggles.

*   Pick the ball up off the floor if it falls to prevent a slip, trip, or fall hazard.

FIGURE 7.5: Celts

SAFETY NOTE

Wear safety glasses or goggles.

RATTLEBACK

A celt (Figure 7.5), sometimes known as a rattleback, seems to be a simple gizmo. However, the physics behind the phenomenon it exhibits is a bit complicated.

Procedure

1. Spin the celt clockwise. Observe.

2. Spin the celt counterclockwise. Observe.

Questions

*   What did you see?

*   Suggest a possible explanation for the phenomenon you observed.

▼
SAFETY NOTE

Wear safety glasses or goggles.

FUN FLY STICK

The Fun Fly Stick (Figure 7.6) is like a small Van de Graaff generator. The generator in the stick creates a positive electric charge on the outer shell of the Fun Fly Stick. The outer shell is made of paper, an insulator. When the stick is charged, you can perform many demonstrations related to static electricity.

Procedure

1. Carefully take out one of the aluminum foil figures that came with the Fun Fly Stick. Put it in one hand. Turn the apparatus on by pushing the button, and charge the foil by touching it with the wand. Carefully shake the Fun Fly Stick to detach the foil figure. After you have charged the foil, you can make it fly as you wish.

2. Next, take the foil figure in the shape of a butterfly. Charge it and make it fly. Put your hand above the flying butterfly so that the butterfly is between your hand and the Fun Fly Stick. Press the button and make the butterfly bounce between your hand and the Fun Fly Stick.

3. Take a sheet of paper, rub the Fun Fly Stick on it against a wall (as though you were ironing it), and make the paper stick to the wall. (This works better on some walls than others.)

Questions

*   How does the Fun Fly Stick work?

*   Can you get an electric shock from the Fun Fly Stick? Why or why not?

*   Why does the aluminum foil figure change shape when it is charged?

*   What makes the figure go up in the air?

*   Why does the butterfly bounce between your hand and the Fun Fly Stick?

*   Why does the paper stick to the wall?

FIGURE 7.6: Fun Fly Stick

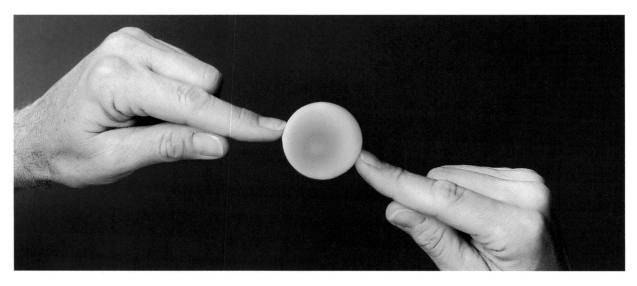

FIGURE 7.7: Energy Ball

ENERGY BALL

The Energy Ball (Figure 7.7) lets you demonstrate some concepts of open and closed electrical circuits.

Procedure

1. Make a circle of students in the classroom. Students must touch each other (e.g., by holding hands or touching elbows).

2. Put the Energy Ball between two students in the circle so that each student touches one of the two metal strips on the Energy Ball.

3. Ask one student in the circle to open the circuit.

Questions

*   How big a circle of students can you form and still make the Energy Ball work?

*   Explain the concepts (a) *open electrical circuit* and (b) *closed electrical circuit*.

*   Find out how the Energy Ball works.

LEVITRON

The physics behind the Levitron (Figure 7.8) is complicated. The magnetic field is formed by a toroidal (doughnut-shaped) magnet in the device. It takes time to get it working— you will have to adjust the weight until it is just right—but once it is working, it is most impressive.

Procedure

1. Adjust the Levitron so that the platform is precisely horizontal.

2. Place the transparent layer on the Levitron platform.

3. Spin the top on the transparent layer and slowly lift the layer about 5 cm above the level of the platform.

4. Slowly lower the transparent layer to the platform, leaving the top levitating.

   - You will need to adjust the mass of the top through repeated trials to get it to levitate. If the top falls directly after being raised, it is too heavy.

   - If the spinning top flies off, the top is too light or the Levitron platform is not level.

5. When you get the top to stay levitated, do the following tests:

   - Move items over, under, and around the top as it rotates (e.g., a sheet of paper or your fingers).

   - Place a drinking glass around the spinning top.

   - Pretend to cut the air above the top with a pair of metal scissors.

FIGURE 7.8: Levitron

Questions

- Why does the spinning top stay levitated?

- How does the paper, a glass, or your finger influence the top?

- What happens when you cut the air with metal scissors? Why?

SAFETY NOTE

Wear safety glasses or goggles.

7A Analysis

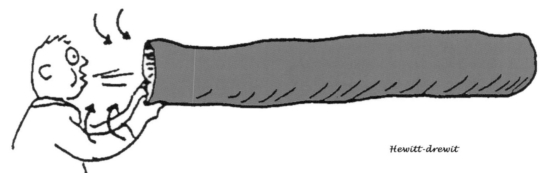

FIGURE 7.9: "Bernoulli's Bag"

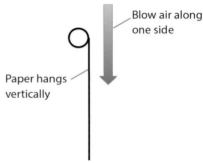

Blow air along one side

Paper hangs vertically

FIGURE 7.10: An experiment showing that moving air does not necessarily have lower pressure

WIND BAG

After the contestants have had their competition with the "Bernoulli's Bags" (Figure 7.9), it's the referee's turn. Hold the bag wide open about 20–30 centimeters from your mouth and blow vigorously once or twice. This will be enough to fill the bag.

Sometimes it is said, based on Bernoulli's principle, that moving air has a lower pressure than stationary air. However, there's a simple experiment you can do to show that simply blowing air out from your lungs does not produce a stream of air with a lower pressure: Hold a piece of paper so that it hangs down vertically, and blow down on one side of the paper (Figure 7.10).

Notice that the paper doesn't deflect toward the moving air. In an open environment, any pressure difference goes to zero in practically no time.

However, blowing out air can produce other effects. One of them is called *entrainment*. Entrainment occurs when the surrounding air gets pulled along with the air stream. Some fans work on this principle to blow out more air than you would otherwise expect. Filling the bag with just one blow can be explained by considering molecules and entrainment. The air molecules close to the stream are carried with the air molecules in the

stream through entrainment. There is much more air coming from outside the stream than in the stream, which causes the bag to fill with a single breath.

How is entrainment different from the Bernoulli effect? The Bernoulli effect involves changes of speed and pressure along flow lines in a tube or other confined enclosure. As mentioned above, in an open space any differences in air pressure would immediately push air from the higher-pressure region toward the lower pressure region, eliminating the original pressure difference between the moving and stationary air. Entrainment works on the microscopic level and is different from the Bernoulli effect.

Firefighters may use this phenomenon when removing smoke from buildings. They place fans at a distance from a door or window, making the rate of airflow much higher.

FIGURE 7.11: Mirage

MIRAGE

The image formed in the Mirage (Figure 7.11) is a real image because light rays are reflected from an object and actually reach the image on top of the Mirage. If you put the pig (which comes in the kit) on the lower mirror so that its nose is toward you, you'll notice that in the image the tail is toward you. Light rays reflected from the pig go from the upper mirror to the lower one, where they reflect to the eye of the viewer. A virtual image cannot be seen this way, and the real image could be reflected, for example, to a white screen or a wall. (Note that many websites refer to the image formed by Mirage as a hologram; however, it is not one.)

DROPPER POPPER

Energy is added to the Dropper Popper (Figure 7.12) when you turn down the edges, and that energy is stored in its rubbery material. When you drop the Popper, the stored energy is released as the Popper returns to its original shape. This makes it bounce higher than the height from which it was dropped.

FIGURE 7.12: Dropper Popper

MAGNETIC ACCELERATOR

The Magnetic Accelerator (Figure 7.13) activity explores the conversion of potential energy to other forms of energy. The balls' potential energy is converted into kinetic energy and then into sound energy and, due to resistance forces such as friction and air resistance, to heat. The fourth ball, which is magnetic, accelerates just before its impact with other balls, thereby gaining enough energy to cause the end ball to fly off the end of the ramp. The same occurs with the last ball, for which the magnetic ball is at rest, because the approaching ball accelerates rapidly because of the magnetic force pulling it forward.

The collisions are almost elastic, and the kinetic energy of the ball is transferred to the last ball in line, which is far enough from the magnetized ball to not be noticeably affected by its magnetic field. The last ball has so much energy from the collision that it easily leaves the ramp.

FIGURE 7.13: Magnetic Accelerator

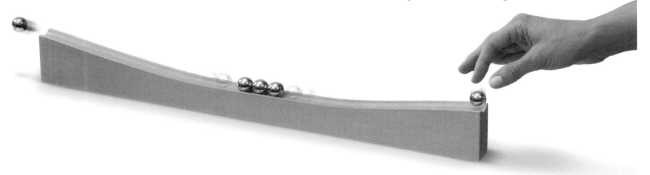

FIGURE 7.14: Celt

RATTLEBACK

The balance point of the celt (Figure 7.14), also known as a rattleback, is not on its axis of rotation. When the celt spins clockwise, friction exerts a force that transforms the rotation to a vertical oscillation. The force caused by the friction is also the reason why the vertical oscillation then transforms back to a rotation—in the opposite direction. If you spin the celt clockwise, its rotational direction changes. If you spin it counterclockwise, it rotates normally and does not change direction.

FUN FLY STICK

The Fun Fly Stick (Figure 7.15) shows that some materials can behave somewhat differently from normal when they come in contact with certain other materials. Some materials tend to donate electrons to other materials and are more likely to get positively charged, whereas other materials tend to take electrons when in contact and receive a negative charge. You may have experienced this when combing your hair. The comb can remove electrons from your hair, leaving the hair positively charged. Since like charges repel one another, your hair might stand on end as the positive charges try to get as far apart from each other as possible. The *triboelectric series* can tell you if some material will charge negatively or positively.

In the Fun Fly Stick, there is a rubber belt that goes around two cylinders. One is made of Teflon (PTFE) and the other is made of metal. When the rubber belt moves around and contacts the Teflon cylinder, electrons get transferred from the rubber band to the cylinder. This leaves the rubber band positively charged. As the positively charged belt

passes over the top metal pulley, electrons from the top of the stick get pulled onto the rubber belt, leaving the stick with a positive charge.

The Fun Fly Stick is covered with cardboard, which is an insulator, so you don't get an electric shock from the stick. Despite the high voltages present, the resistance of the cardboard prevents the formation of large electrical currents.

When the aluminum foil is in contact with the Fun Fly Stick, free electrons in the foil transfer to the positively charged stick, leaving the foil with an excess of positive charges. Because like charges repel each other, the aluminum foil opens into a large shape. In addition to the repulsion of charges, gravity and air currents affect the foil while it moves through the air.

In one step, you bounce the butterfly between your hand and the Fun Fly Stick. This happens because after the butterfly has been charged, the negative charges in your hand attract the foil butterfly. When the butterfly touches your hand,

FIGURE 7.15: Fun Fly Stick

some electrons get transferred, and then the hand's contact point is positively charged so that the like charges repel each other and the foil butterfly moves away from your hand. The positively charged Fun Fly Stick then attracts the free electrons from the foil and the process starts over.

In the final demonstration of the power of static electricity, you stick a piece of paper to the wall. Paper is an insulator, so it does not have many free electrons as charge carriers. Sliding the Fun Fly Stick along the paper can do two things to the paper: First, it can polarize some of the molecules in the paper. The molecules on the paper's surface tend to turn so that their negative poles are directed toward the positively charged Fun Fly Stick. Second, it can cause some electrons to move from the paper to the Fun Fly Stick, leaving the paper with a net positive charge. In response to the paper's positive charge, and the positively charged Fun Fly Stick, two things can happen in the wall: One, the molecules in the wall can turn so that their negative poles face the paper and attract it. Two, some electrons will flow through the wall to the region behind the paper. Those excess electrons will also attract the positively charged paper.

ENERGY BALL

The Energy Ball (Figure 7.16) exploration helps you investigate closed and open circuits by creating a human circuit. When the circuit is closed (i.e., everyone is holding hands or touching elbows), the ball makes noise and flashes a light.

The voltage of a battery causes an electric current in a closed circuit, which is a loop that enables the current to return to its starting point. In the circuit, there is usually a source of voltage and current, resistors, and sometimes capacitors or inductors. In an open circuit there is no current present

FIGURE 7.16: Energy Ball

in spite of having voltage. A closed circuit can be made open (and vice versa) with a switch.

The operation of the Energy Ball is based on two different circuits. In the Energy Ball there is a channel field-effect transistor through which the current passes if the resistance between the metal strips is measurable. In the circuit made by students, there is a very small current. When that current is flowing, the transistor allows current in another circuit to flow through the lamp and speaker.

LEVITRON

Here, you spin a top and then watch as (after a certain amount of tinkering) it levitates above a platform. When the top is spinning above the magnet in the Levitron's base (Figure 7.17), the top is affected by magnetic repulsion, gravity, and air resistance. Because the magnetic repulsion is as strong as gravity, the spinning top is levitating in the Levitron's magnetic field. When the top's spinning slows down, it falls. The most significant force to slow down the top is air resistance.

FIGURE 7.17: Levitron

If scissors are made from a ferromagnetic material, such as steel, they will be magnetized and will cause enough of a disturbance to the Levitron's magnetic field to make the top fall because of gravity. Paper, a glass, or fingers don't affect the magnetic field.

Web Resources

More information about the Mirage device.
www.arborsci.com/media/datasheet/P2-7070_Additional.pdf

Aerodynamic lift explained (and some comments about the wind bag near the end).
www.aerodynamiclift.com

An article about Dropper Poppers.
www.nytimes.com/2001/12/09/magazine/the-year-in-ideas-a-to-z-dropper-popper.html

A video of the Mirage device.
http://physics.wfu.edu/demolabs/demos/6/6a/6A2035.html

A Wikipedia article about celts.
http://en.wikipedia.org/wiki/Rattleback

More information about the Fun Fly Stick, including a short list of the triboelectric series.
www.unitechtoys.com/pdf/FunFlyStickEducationalWeb.pdf

One-minute video of Energy Ball.
www.youtube.com/watch?v=pgT_9a5jMqM

Discussion on the use of the Magnetic Accelerator.
www.buzzillions.com/reviews/arbor-scientific-magnetic-accelerator-reviews

The physics of the Levitron.
www.levitron.com/physics.html

Relevant Standards

Note: The Next Generation Science Standards *can be viewed online at* www.nextgenscience.org/next-generation-science-standards.

WIND BAG

Disciplinary Core Ideas

PS1.A: Structure and Properties of Matter

Gases and liquids are made of molecules or inert atoms that are moving about relative to each other. (MS-PS1-4)

MIRAGE

Disciplinary Core Ideas

PS4.B: Electromagnetic Radiation

- When light shines on an object, it is reflected, absorbed, or transmitted through the object, depending on the object's material and the frequency (color) of the light. (MS-PS4-2)

- The path that light travels can be traced as straight lines, except at surfaces between different transparent materials (e.g., air and water, air and glass) where the light path bends. (MS-PS4-2)

DROPPER POPPER

Performance Expectations

MS-PS3-5

Construct, use, and present arguments to support the claim that when the kinetic energy of an object changes, energy is transferred to or from the object. [Clarification Statement: Examples of empirical evidence used in arguments could include an inventory or other representation of the energy before and after the transfer in the form of temperature changes or motion of object.] [Assessment Boundary: Assessment does not include calculations of energy.]

MAGNETIC ACCELERATOR

Performance Expectations

MS-PS2-5

Conduct an investigation and evaluate the experimental design to provide evidence that fields exist between objects exerting forces on each other even though the objects are not in contact. [Clarification Statement: Examples of this phenomenon could include the interactions of magnets, electrically-charged strips of tape, and electrically-charged pith balls. Examples of investigations could include first-hand experiences or simulations.] [Assessment Boundary: Assessment is limited to electric and magnetic fields. Assessment is limited to qualitative evidence for the existence of fields.]

MS-PS3-2

Develop a model to describe that when the arrangement of objects interacting at a distance changes, different amounts of potential energy are stored in the system. [Clarification Statement: Emphasis is on relative amounts of potential energy, not on calculations of potential energy. Examples of objects within systems interacting at varying distances could include: the Earth and either a roller coaster cart at varying positions on a hill or objects at varying heights on shelves, changing the direction/orientation of a magnet, and a balloon with static electrical charge being brought closer to a classmate's hair. Examples of models could include representations, diagrams, pictures, and written descriptions of systems.] [Assessment Boundary: Assessment is limited to two objects and electric, magnetic, and gravitational interactions.]

Disciplinary Core Ideas

PS2.B: Types of Interactions

- Electric and magnetic (electromagnetic) forces can be attractive or repulsive, and their sizes depend on the magnitudes of the charges, currents, or magnetic strengths involved and on the distances between the interacting objects. (MS-PS2-3)

RATTLEBACK

Science and Engineering Practices

Constructing Explanations and Designing Solutions

Constructing explanations and designing solutions in 6–8 builds on K–5 experiences and progresses to include constructing explanations and designing solutions supported by multiple sources of evidence consistent with scientific ideas, principles, and theories.

- Apply scientific ideas or principles to design an object, tool, process or system. (MS-PS2-1)

Engaging in Argument From Evidence

Engaging in argument from evidence in 6–8 builds from K–5 experiences and progresses to constructing a convincing argument that supports or refutes claims for either explanations or solutions about the natural and designed world.

- Construct and present oral and written arguments supported by empirical evidence and scientific reasoning to support or refute an explanation or a model for a phenomenon or a solution to a problem. (MS-PS2-4)

FUN FLY STICK

Performance Expectations

MS-PS2-5

Conduct an investigation and evaluate the experimental design to provide evidence that fields exist between objects exerting forces on each other even though the objects are not in contact. [Clarification Statement: Examples of this phenomenon could include the interactions of magnets, electrically-charged strips of tape, and electrically-charged pith balls. Examples of investigations could include first-hand experiences or simulations.] [Assessment Boundary: Assessment is limited to electric and magnetic fields. Assessment is limited to qualitative evidence for the existence of fields.]

MS-PS3-2

Develop a model to describe that when the arrangement of objects interacting at a distance changes, different amounts of potential energy are stored in the system.

[Clarification Statement: Emphasis is on relative amounts of potential energy, not on calculations of potential energy. Examples of objects within systems interacting at varying distances could include: the Earth and either a roller coaster cart at varying positions on a hill or objects at varying heights on shelves, changing the direction/orientation of a magnet, and a balloon with static electrical charge being brought closer to a classmate's hair. Examples of models could include representations, diagrams, pictures, and written descriptions of systems.] [Assessment Boundary: Assessment is limited to two objects and electric, magnetic, and gravitational interactions.]

ENERGY BALL

Crosscutting Concepts

Energy and Matter

The transfer of energy can be tracked as energy flows through a designed or natural system. (MS-PS1-6)

APPENDIX

HOW TO ORDER THE GADGETS AND GIZMOS

Materials to support the lessons and experiments in this book are available from Arbor Scientific in the NSTA Middle School Physics Kit. The kit includes about 36 gadgets to support lessons in thermodynamics, pressure, energy, waves, buoyancy, and more.

Visit *arborsci.com* to order the kits or learn more about their contents.

Kit #PK-0300

Chapter	Gadget
1. Wave Motion and Sound	tuning forks (256, 384, 426.7, 288, 320 Hz)
	5 - Standing Wave Apparatus
	5 - Sound Pipe
	Music Box
	Doppler Ball
2. Visible Light and Colors	Giant Prism
	5 - Spectroscope
	Primary Color Light Sticks
	RGB Snap Lights and Spinner
3. Thermodynamics	Radiation Cans
	Ice Melting Blocks
	Ball and Ring
	5 - Drinking Bird
	Fire Syringe
4. Air Pressure	Atmospheric Pressure Mat
	5 - Atmospheric Pressure Cups
	Pressure Globe
	Bottle Rocket Launcher
5. Density and Buoyancy	Steel Sphere Density Kit
	Boat and Rock
	Solar Bag
	5 - Cartesian Diver
6. Force, Motion, and Energy	5 - Pull Back Car
	5 - Balancing Bird
	Inertia Apparatus
	Vertical Acceleration Demonstrator
	Energy and Motion Lab
	5 - Happy/Unhappy Balls
7. It's Science, Not Magic	Bernoulli Bags 4/pk.
	Mirage
	5 - Dropper Popper
	Magnetic Accelerator
	5 - Celt, pkg of 2
	Fun Fly Stick
	5 - Energy Ball
	Levitron w/Starter, Battery Included

KEYWORD GLOSSARY

acceleration The rate of change of velocity. Since velocity involves both speed and direction, any change in speed or direction means that there is acceleration. Acceleration can be calculated by dividing the change in velocity by the change in time, abbreviated $a = \Delta v \,/\, \Delta t$. The metric units of velocity are m/s^2.

air pressure The force per unit area that is exerted by the air on any surface in contact with it. For a horizontal surface in open air, the air pressure is equal to the weight of the air above the surface. Air pressure pushes equally in all directions—not only downwards. The metric unit of air pressure is the pascal (Pa). One pascal is the pressure of a one newton force spread over one square meter of area.

amplitude The extent of the largest disturbance from the undisturbed position in a vibrating system.

atmosphere The 100-kilometer thick layer of air surrounding Earth (or the layer of gas surrounding any celestial body). The Earth's atmosphere helps maintain conditions suitable for life on Earth. It contains 78% nitrogen, 21% oxygen, and small amounts of other gases, such as argon, water vapor, carbon dioxide, and so on. Due to the greenhouse effect, the atmosphere keeps the surface of the Earth warmer than it would otherwise be, and it blocks some of the harmful ultraviolet radiation from the Sun.

Celsius scale A temperature scale in which 0°C is the melting point of water (ice) and 100°C is the boiling point of water. For example room temperature is approximately 20°C, and the human core temperature approximately 37°C. The scale was defined by a scientist named Anders Celsius in the 18th century.

CMYK cyan, magenta, yellow, key (black). A subtractive color model that is mostly used in printing. Cyan, magenta, and yellow are the secondary colors of light—formed by combining two primaries. They are also the primary pigments and are used to form different colors with paints and dyes, in printers, and so on.

density A measure of how compact the matter in an object is. Density is calculated by dividing the mass of an object by its volume. The metric units of density are kg/m^3, or kilograms per cubic meter.

echo A reflection of sound waves. For example, you can hear an echo when you clap your hands in a large, empty room. The sound waves created when clapping hands reflect from the walls, so you are able to hear the sound again.

energy An object's or system's ability to do physical work. Energy can be stored in different systems in different ways. Energy can also be transferred from one system to another by different methods. However, it is all the same thing—energy. The amount of all the energy (including mass-energy) in the universe is constant. It cannot be created or destroyed. The metric unit of energy is the joule. A 100-watt light bulb uses 100 joules of energy each second.

Fahrenheit scale   A temperature scale in which the lower defining point is the lowest temperature that scientist Daniel Gabriel Fahrenheit could reach in his lab in the 18th century. He defined that temperature as 0 degrees Fahrenheit. The upper defining point was the human core temperature, which he chose to be 100 degrees (although it's actually closer to 98.6°F). On the Fahrenheit scale, room temperature is approximately 68°F.

floating   Remaining near the surface of or in a fluid, which can be a liquid or gas, due to the buoyancy from the fluid. An object will float if its density is less than the density of the fluid.

force   A push or a pull on objects or systems due to interactions between them. The amount of force tells how strong the interaction is. The metric unit of force is the newton (N), which is the same as kg·m/s². Unbalanced forces will result in an acceleration of the object. A force can also change the shape of an object. Force can be calculated by dividing the change in an object's momentum by the change in time, abbreviated $F = \Delta(mv) / \Delta t$. Force can also be expressed as the product of an object's mass and its acceleration, or $F = ma$. A force of 1 N will give a 1-kg mass an acceleration of 1 m/s².

frequency   The number of vibrations during a unit time. The metric unit for frequency is hertz, abbreviated Hz. For example 50 Hz means that there are 50 vibrations in one second. The human ear can hear frequencies from 20 Hz to 20,000 Hz.

gravitational force from the Earth on an object   The weight of the object, calculated by the product of the object's mass and the acceleration of gravity, abbreviated $F_g = mg$.

gravitational potential energy   Stored energy due to an object's position in the Earth's (or other body's) gravitational field. This potential energy can later be converted to kinetic energy or used to do work. The gravitational potential energy, E_p, of an object can be calculated by multiplying an object's weight (mg) by its height, h, abbreviated $E_p = mgh$.

heat   Thermal energy moving between two substances due to their difference in temperature.

Kelvin scale   A temperature scale (see *thermometer*) in which the lower defining point is absolute zero, 0 K, and the upper is the "triple point" of water, which is the particular temperature and pressure at which solid, liquid, and gaseous water can all exist together. The temperature of the triple point of water is defined to be 273.16 K (0.01°C). The Kelvin scale has the same size degrees as the Celsius scale, but Kelvin readings are 273.16 degrees higher. Room temperature (20°C) is 293.15 K. The Kelvin scale was defined by William Thomson, also known as Lord Kelvin, in the 19th century. The Kelvin temperature scale is widely used in science.

kinetic energy   The energy of an object due to its motion. Kinetic energy, E_k, can be calculated by the formula, $E_k = \frac{1}{2} mv^2$. The metric unit of energy is the joule.

law of conservation of energy   The principle that energy (including mass-energy) can be transformed from one form to another but cannot be created or destroyed. Therefore, in an isolated system, the starting energy is equal to the final energy, no matter what changes take place within the system. $E(\text{initial}) = E(\text{final})$

mass   The amount of matter in an object. The mass of an object is also a measure of the inertia of the object. The greater the mass, the greater the inertia, and the lower the acceleration of an object in response to a force. The metric unit of mass is the kilogram.

medium   A substance through which a wave can be carried. For example, a medium for sound waves could be air, water, or metals (but not a vacuum).

prism   In science, a prism is a transparent object, often glass, usually in a triangular shape, that refracts or reflects light. (See *refraction of light* and *spectrum*.) Because different colors refract at different angles, a prism can be used to break white light up into a spectrum of colors. A prism can also be used to demonstrate total internal reflection.

reflection of light   Light waves change their direction when they encounter the boundary between two media. When reflecting, the waves come back to the original medium. For example, you can see a reflection in a mirror or on the surface of water.

refraction of light   Refraction is a bending of a wave as it crosses the boundary between two different media. For example when light travels at some angle from air into water it refracts. That means that its direction changes.

resonance   The natural vibration frequency of a system. A system oscillates at greater amplitudes at resonant frequencies. For example, in musical instruments some frequencies—the resonant frequencies—are naturally louder than others.

RGB   red, green, blue. RGB represents an additive color model. As the three colors are combined in different amounts, we can produce a huge variety of different colors. The RGB color model is used in different kinds of display screens, such as those in laptops, mobile phones, and so on.

sinking   Descending into a fluid, which occurs if an object's density is greater than the density of the fluid.

sound wave   Vibration that travels through a medium. Sound waves in air consist of regions of air compressions and rarefactions. Sound waves are longitudinal waves created by vibration. In solid materials, sound can travel as both longitudinal and transverse waves.

spectrum   The common name for a phenomenon in which a physical quantity is displayed as function of frequency or wavelength. One familiar example of a spectrum is a rainbow. In rainbows, the white light from the Sun refracts in water drops to create a circular spectrum. (See *refraction of light*.)

thermometer   A device used to measure temperature. Traditional thermometers work because of the thermal expansion of a liquid inside a tube; however there are other kinds of thermometers as well. See *Celsius scale*, *Fahrenheit scale*, and *Kelvin scale*.

vacuum   A space that does not contain any matter. A perfect vacuum, although not actually possible to attain, would have a pressure in it of zero. A partial vacuum has some pressure, but less than normal atmospheric pressure. A vacuum is created by removing air from a container. A vacuum cleaner works by creating a partial vacuum inside the machine, and then the greater outside air pressure forces air into it, dragging dirt along with the air.

velocity   The speed and direction of motion of an object. An object's speed is how many meters (or any length unit) an object moves per second (or any time unit). The average speed of an object moving can be calculated by dividing the distance traveled by the time, or $v = d/t$. The metric unit of velocity is m/s—meters per second.

volume   The amount of space that an object or substance occupies. The metric unit of volume is m^3, pronounced "cubic meters."

wave motion   Movement of energy from one place to another without moving particles or matter from place to place. A vibrating object can create wave motion as the vibrations travel through a medium. Sound is an example of wave motion.

wavelength   The length of a wave. The wavelength is the distance between two successive crests of a wave, or the distance between any two consecutive points that are at the same phase. In the metric system, wavelength is measured in meters.

weight   The amount of force that gravity exerts on an object. The weight of an object is equal to the product of the object's mass and the acceleration of gravity, abbreviated $F_g = mg$. Since weight is a force, it can be expressed with the metric unit of newtons.

work   The transfer of energy from one object or system to another as the result of a force. For a constant force, the work can be calculated by the product of the force and the distance over which the force was applied, or $W = Fd$. The units of work are the same as for energy, so work can be expressed in units of joules.

CREDITS

IMAGES

MEDIAKETTU JARI PEURAJÄRVI

ARBOR SCIENTIFIC

RPM SPORTS

PAUL HEWITT

MATTI KORHONEN

OTHER CONTRIBUTIONS

OLIVIA BOBROWSKY

ANDREW GLOR

MARIE BRETON

INDEX

*Page numbers in **boldface** type refer to tables or figures.*

INDEX

INDEX

Drinking Bird, **31, 36**
Fire Syringe, **32, 37**
Ice Melting Blocks, **29, 34**
Radiation Cans, **28, 33**
for Visible Color and Light explorations
Giant Prism, **18, 22**
Primary Color Light Sticks, **20, 23**
Quantitative Spectroscope, **19, 22**
RGB Spinner and Snap Lights, **21**
for Wave Motion and Sound explorations
Doppler Ball, **6**
Music Box, **5, 10**
Sound Pipe, **4, 8**
Standing Wave Apparatus, **3, 7**
Tuning Fork, **2, 7**
Gay-Lussac's law, 37
Get in Tune exploration, 2, **2**
analysis of, 7, **7**
Glossary, 105–108
Gravitational force, 60, 61, 76, 106
Gravitational potential energy, 67, 77, 106

H
Happy/Unhappy Balls exploration, 73, **73**
analysis of, 78, **78**
Harmonics, 8, **8, 9**
Hearing, 1
Heat, 27, 106
Heat conduction, 33, 34
Hydrostatic pressure, 62

I
Index of refraction, 22
International System of Units (SI units), 43
It's a Hold-Up! exploration, 44, **44**
analysis of, 48, **48**
It's a Snap! exploration, 70, **70**
analysis of, 75, **75**
It's Science, Not Magic explorations, 85–102
Dropper Popper, 88, **88**, 95, **95**
Energy Ball, 92, **92**, 97, **97**
Fun Fly Stick, 91, **91, 96,** 96–97
Levitron, 93, **93**, 97, **97**
Magnetic Accelerator, 89, **89**, 95, **95**

Mirage, 87, **87**, 95, **95**
Rattleback, 90, **90**, 96, **96**
standards addressed by, 99–102
Web resources for, 98
Wind Bag, 86, **86,** 94, **94**

J
Journaling, xiii

K
K-W-L approach, xii
Kelvin scale, 27, 106
Kinetic energy, 67, 72, 73, 106
during collisions, 37, 95
conversion of potential energy to, 67, 77, 78, 95, 106
formula for, 67, 106
mass, velocity and, 77
Kits of gadgets and gizmos, 103

L
Law of conservation of energy, 67, 106
Law of conservation of energy explorations
Dropper Popper, 88
Energy Ingenuity, 72, 77
Happy/Unhappy Balls, 73
Magnetic Accelerator, 89
Learning goals, xii–xiii
Levitron exploration, 93, **93**
analysis of, 97, **97**
Light. *See also* Visible Light and Colors explorations
reflection of, 17, 23, 33, 95, 107
refraction of, 17, 18, 22, 107
Longitudinal (compression) waves, 1, 107

M
Magnetic Accelerator exploration, 89, **89**
analysis of, 95, **95**
standards addressed by, 100
Mass, 55, 107
Medium for sound waves, 1, 107, 108